THE PRODIGIOUS BUILDERS

Books by Bernard Rudofsky

Are Clothes Modern?
Behind the Picture Window
Architecture Without Architects
The Kimono Mind
Streets for People
The Unfashionable Human Body

Book design by the author

Bernard Rudofsky

The Prodigious Builders

notes toward a natural history of architecture with special regard to those species that are traditionally neglected or downright ignored

Harcourt Brace Jovanovich New York and London

Printed in the United States of America

Library of Congress Cataloging in Publication Data
Rudofsky, Bernard, 1905–
The Prodigious Builders.
Includes bibliographical references.
1. Architecture. I. Title
NA200.R76 720′.9 77-73124

ISBN 0-15-193050-3

First edition

B C D E

Parts of the chapters "In Praise of Caves" and "Strongholds" first appeared in somewhat different form and under different titles in *Horizon* magazine, © 1967 and 1971 by American Heritage Publication Co., Inc.

These notes toward a natural history of architecture resume arguments broached in *Architecture Without Architects*, an earlier book on non-pedigreed architecture. They deal with architecture as a tangible expression of a way of life rather than as the art of building. Moreover, the material at hand is presented from the naturalist's point of view as distinct from that of the historian. The digressions and detours in the text are intended to break down the barriers that separate us from the builders of alien and archaic architecture, people who never needed to be told what was good for them. Their unpretentious and occasionally awesome achievements merit consideration; no architecture, it seems to me, is outdated that works for man rather than against him.

Contents

Introduction with asides

> That was a happy age, before the days of architects, before the days of builders.
>
> SENECA (c. 4 B.C.–A.D. 65)

In its entirety this curiously modern-sounding passage from Lucius Annaeus Seneca's letter to Lucilius reads: "Life is the gift of the immortal gods, living well is the gift of philosophy. Was it philosophy that erected all the towering tenements, so dangerous to the persons who dwell in them? Believe me, that was a happy age, before the days of architects, before the days of builders."[1] With a classical flair for economy, Seneca managed to compress a pungent aphorism, a barbed rhetorical question, and a poignant valedictory into three short sentences. They will recur, slightly veiled, as leitmotifs in this essay on some of the lesser-known aspects of architecture.

Architecture without architects, as I call the topic at hand, is not just a jumble of building types traditionally slighted or altogether ignored, but the silent testimonial to ways of life that are heavy on acute insight, albeit light on progress. It goes to the roots of human experience and is thus of more than technical and aesthetic interest. Moreover, it is architecture without a dogma.

In our days, when it has become permissible, indeed proper, to blame progress for about every evil that has befallen this planet, it is a comforting thought that, everything considered, we have not much progressed in those disciplines that proclaim most emphatically humanity's humaneness: poetry, music, and the arts. With them, change has been negligible; our hearts still beat to the old rhythms. We are worshipping—after a fashion—artists and thinkers of distant ages whose work, or fragments of it, have come down to us and are still capable of stirring our blood. Come to think of it, the ancients may very well have been inclined to return the compliment. It is perfectly conceivable that Ovid might have enjoyed reading Eliot in a good translation; that audiences of, say, Aeschylus' *Agamemnon* might have been amenable to Orff's unpolished subtleties, or that Miró, after some initial misunderstandings, might have been acclaimed as a brother-in-arms by all underground artists from Lascaux to the Roman catacombs.

Architecture represents a special case. At its best, and not necessarily its loftiest, it is, or was, and surely ought to be, all three: art, poetry, and music—*not* what Goethe rather inaptly called "frozen music" but a medium that, when it transcends utility, can be best described in musical terms. As to poetry, there was never any dearth of dirges on the death of a building or a town, or of elegiacal musings about the flawless beauty of architecture's ruinous, ivy-infested stages. However, as architecture proliferated, it lost its integrity. At one point some of it succumbed to perpetual progress, never to recover. Just like the fateful split of Noah's stock resulted in three incongruous tribes, architecture was torn asunder into three irreconcilable spheres—one staked out by the dumb but resourceful animals; another by the various stick-in-the-mud factions, represented by prehistoric and unhistoric architecture; and the third by that progressive-aggressive profession whose unattractive products are uppermost in our mind, if only for their inherent uppishness. It is the second, least familiar, species that we will be mostly concerned with.

The teachings of architecture as conducted under academic auspices leave little or no room for studying undatable monuments. They rank as low as wines *sans année*. Orthodox architectural history—that social register of more or less grandiloquent buildings, closely related to each other by family ties but isolated from the vast mass of anonymous architecture—exhilarating as it is when dispensed by an imaginative teacher, more often than not turns into a dispirited romp through the centuries and gives only an incomplete account of man as a builder. Worse, it cheats us of the beginnings. It skips the bucolic and heroic stages of architecture's small hours, and presents instead a bewildering catalog of monumental minutiae, generously asterisked and dragging the ball and chain of footnotes. Yet pedigreed architecture, from Babylon to Brasilia, does not add up to anything near a complete record of man's building enterprises. Its poorest relation, prehistoric architecture, even when seen through the wrong end of the telescope, appears to be a storehouse of human experience.

Make no mistake; that noble ancestor, prehistoric man, whom we vaguely imagine through the hazy past as a kind of postgraduate primate, was quite a bright fellow. He probably had a higher I.Q. than our neighbor next door; his brain certainly was bigger. He may have had the olfactory equipment of a sporting dog, and 20-20 vision far into old age, if only because he never read small print by poor light. A big-game hunter by necessity, he always found time for painting and sculpting on a big scale, and when he played with boulders, the results—architectural birthmarks rather than buildings—were often spectacular. (Churchill's flair for bricklaying is a case of atavistic pandering to this hoary and gentlemanly pastime.) Prehistoric man was, to be sure, technically illiterate. He had no words for expressionism or symbolism, yet in matters of the plastic arts he possessed expertness infused by an absentminded sort of genius. In a way he had more practical wisdom than modern man, for what we call his "primitive" dwellings were dwellings governed by ecological factors. That he was generous we may surmise from the scale of his work. Long before building materials were miniaturized into bricks and ashlars, he accomplished feats that command respect. In all his constructions he was moved by a sense of grandeur. If he has been misjudged and ridiculed, we ourselves are to blame.

People adverse to ruminating upon Stone Age remnants may find it more stimulating to turn their thoughts to that live though sometimes no less primitive architecture of nations that only yesteryear were labeled and libeled savages. Through all but one-twentieth of humanity's life span man led the life of a savage, which means that he subsisted on a so-called wild-food economy, gathering shellfish, plants, fruit, and the like. Paradoxically, his habitation was conceived and contrived with far more imagination than that of today's urban savage. Anthropologists have long counted among his admirers, and lately, squads of bright young architects, keen on extracurricular adventure, have taken to foraging those unfamiliar dwellings for proof of built-in lessons of architectural *savoir-faire*. The very same huts that once lent a touch of vaudeville to our World's Fairs, tenanted as they were by snake charmers and belly dancers, become authentic works of art when their merits are vouched for and their contents perused by such worthies as, for instance, André Gide (p. 271). Even a nodding acquaintance with these exotic habitations could help, if not exactly to tear down, at least to jolt, the layman's inveterate misconceptions of domestic architecture. Of course, merely to describe and picture overly foreign dwellings to the reader without being able to offer him an opportunity to test their amenities is about as satisfactory as reading to him the recipes for *ortolans à la toulousaine* or *ponty füszermártással* without letting him taste these dishes.

On one point no consensus can be expected—the definition of what makes shelter fit for humans. It is one of the topics at the bottom of our inquiry. Stone Age man might have wondered what we see in that vulnerable commodity we call Home. High winds and earthquakes are apt to reduce it to kindling. Floods sometimes mistake it for the Ark. Housemovers will obligingly pull it to another location. Yet the house looms large, if not as a refuge, as a metaphor, live, dead, and mixed. It is the repository of our wishes and dreams, memories and illusions. It is, or at least ought to be, instrumental in the transition from being to well-being. Bricklayer Churchill sized up architecture as a mold for the human mind when he said, "We shape our buildings, and afterwards our buildings shape us."

Clearly, architecture means more than a roof over one's head and the constraint of walls. There was a time when men got together for no other purpose than to put up mile-long processions of good-for-nothing but enchanting monoliths (p. 109, fig. 71); or to compose a stone circle, that elementary roundelay at which the inspired chimpanzee arrives without example or instructions (p. 107, fig. 70). In the remote past, now dimmed to a flicker of racial memory, piling stone upon stone was neither a trade nor an art but more likely the consequence of an irrepressible urge. While some of the know-how that went with it transpired to people who kept an ear close to the ground, it remains obstinately alien to the city slicker. To be sure, architecture's emotional powers are inferior to those of music. Occasionally, it may have poetic depth, but there is no inherent passion. It causes no sexual arousal. At any rate, no legends are spun around the architecture of our days. Poets and painters abstain from celebrating the White House or Buckingham Palace, buildings of no mean importance in the life of the nations but low on spiritual assets. At best they figure on postal stamps, but no artist worth his salt wants to have anything to do with them.

The inelegant definition of architecture, perpetuated by pedants, as "the art of so building as to apply both beauty and utility,"[2] obviously ought to be expanded to include that vast number of the least ornate of species, lumped together collectively as anonymous architecture. On the face of it it seems preposterous that the vernacular, the very passkey to the understanding of alien cultures, should be so wilfully ignored. Such prejudice does not redound to our honor—would we call botany a science if it dealt exclusively with lilies and roses? Perhaps because it comes mostly in small denominations, such as huts and homely houses, the vernacular fails to enlist respect. Yet one would think that its enormous variety alone, comparable to the variety of biological forms, ought to make it a subject of wide interest. To students of architecture, especially, it is invaluable as therapeutic irritant.

Vernacular architecture owes its spectacular longevity to a constant redistribution of hard-won knowledge, channeled into quasi-instinctive reactions to the outer world. So-called primitive peoples have none of the devil-may-care attitude when confronted with the reality of their environment. Above all, they have no desire to dominate it. Admittedly, the vernacular's unforgivable weakness is constancy. Unlike the apparel arts or pedigreed architecture, it follows no fads and fashions but evolves only imperceptibly in time. As a rule, it is tailored to human dimensions and human needs, without frills, without the hysterics of the designer. Once a life style has been established and habit has begotten a habitation, change for change's sake is shunned.

In some places the exclusive reliance on local building materials alone guarantees the persistence of time-honored construction methods. Conversely, when alien materials and alien methods are introduced, local traditions wither away, customs are displaced by trends, and the vernacular perishes. The question is not whether the demise of the vernacular is merely regrettable—most people couldn't care less—but whether life in general is impoverished by it. To put it differently: Does the disappearance of architectural species native to the soil upset the balance of civilizations in the same way as the disappearance of certain animals and plants upsets the ecological balance?

Today's young are never told that they were born into a shrinking world in the sense that the sum total of things that have become extinct is greater than that of those that exist today. And this is not just a matter of wildlife or raw materials. The most spectacular inroads are those made into man's intellectual exercises; our schools of higher learning are retailing training instead of education. Irony will have it that during the last generation alone any number of crafts have disappeared. With today's know-how we are no doubt able to cultivate peanuts on Mars, but nobody knows, or cares to know, how to make shoes for human feet.

Not that anybody mourns the loss of the old crafts or of hoary building techniques—certainly not the student of architecture who has to regurgitate the ill-digested fruits of learning on examination days (I am referring to bona fide schools). He will tell you that far too much has been made of mankind's architectural idiosyncrasies. To his chagrin, he is forever guided, mostly on hand of photographs, to masterpieces succulent with history, to the detriment of lesser architecture. The plain and the plain bad are rarely up for discussion. He never reaches those heights of detachment from which he can clearly see architecture not just as shelter but as so many containers and conditioners of different ways of life. Perhaps what he needs most is a Pathology of Architecture. It might benefit him no less than those admonitory handbooks on venereal diseases which enlightened parents press on their children.

Cultural disparities are a formidable stumbling block in evaluating alien institutions. At the bottom of all our misunderstandings over the visual arts, and architecture in particular, lies the lamentable assumption that two or more people looking at the same object, or the same view, see the same thing. As a matter of fact, they don't. They would not be human if they did. And yet, all our so-called art education and art appreciation, couched in the conventions of the day, aim more or less at standardizing our sensory impressions.

Our mental attitude toward the world around us is acquired, almost unconsciously, years before we are capable of forming opinions and making judgments of our own. And it is not philosophers who have a hand in shaping our fortitude but their traditional antagonists, the clergymen. Paradoxically, the architectural facts of life are imparted in the course of religious instruction. The bread and wine of the Holy Scriptures, suitably sugared and diluted when dispensed to the child, provide—apart from succor and a cheerless view of life—a good deal of disorientation in the provinces of ethics and aesthetics. The surrealist stories of Noah's Ark (the first mobile dwelling) and the Tower of Babel (the first high-rise) become as familiar to the child as Cinderella and Snow White. Like these sadistic and necrophiliac fairy tales, the religious parables seethe with ambiguities and hidden meanings. The very dénouement raises a host of questions, left forever dangling. Curiously, the holy book has nothing to say about Adam's house. Although it gives a comprehensive account of the bungled first clothes, Adam's attempts at building a shelter are not mentioned.

Of course, Adam, destined as he was for a life of glorified sloth, never amounted to a doer; work ethic had no place in the newly created world. Besides, the Lord had a poor opinion of work which, characteristically, he inflicted upon the First Parents as punishment and curse. Poor Adam had missed out on all those experiences that go into the shaping of desirable human traits. He had skipped infancy and childhood. He had never gone through adolescence with its attendant bodily and intellectual development. He sprang to life an Instant Man. Moored as he was in an unnervingly lush haven with its air of unreality about it that stymied initiative and assertiveness, he could hardly be expected to direct his thoughts toward a shelter for which he had no need. The Fall changed all this. Already the second generation brought forth ruthless Cain, city planner and murderer.

Unfortunately, the chroniclers were not interested in the art of building, nor do they seem to have been endowed with the sort of insight that lifts the reporter to the level of historian. Thus, they expect us to put up with the unlikely story of Enoch's town—the Old Testament's foremost architectural anachronism—without bothering to explain where all the masons, carpenters, and assorted artisans came from who worked on the monumental enterprise at a time when the world's entire population consisted of just one family. Could the mysterious workers have been moonlighting angels?

Genesis' other architectural parable par excellence is the misadventure connected with some unwise men's attempt to build a skyscraper. Hindsight apart, it is safe to assume that, given their modest technological know-how, the wayward Babylonians never had a chance of reaching Heaven and aesthetic bliss by means of an all-brick structure. Neither did the Deity have cause to worry about their ambitious project. Omniscience precluded any serious doubts about its outcome. Rather it may have been a case of jealousy—Jehovah's pique over the popularity of his rival Marduk, the magicians' god. Or should we simply read into the Scriptures' account of the Tower a universally valid warning that an excess of building zeal may have fatal consequences?

The grandiose scheme fell through, but the mania for building towers has not abated with time. On the contrary; at a distance of a hundred generations and thousands of miles from the plain of Shinar, we erect architectural monstrosities that scrape, if not Heaven, the dirty clouds. Yet ours is neither a case of blasphemy nor of all-out greed; our craze for towering buildings goes beyond the desire for glory or getting rich. It seems to be a phenomenon as uncontrollable as the erection of man's copulatory organ.

The tower's opposite, the cave—a uterine symbol according to the doctors of the soul—has been widely used as human shelter. It is one of the subjects dealt with at length in these pages. And so will be tents. Another chapter is reserved for the architecture of beasts. They, not humans, were the first builders; man is but a late-latecomer. If we set a generous figure for his tenancy, it amounts to no more than a couple of hundred thousand years, a mere instant compared to that of the animals. Our planet was, for what seems like eternity, constitutionally, an animal kingdom.

Fate charitably conceals from us the face of things to come, thereby preserving our sanity. Romulus and Remus, taking in the splendors of the twenty-two Romes in the United States; Columbus, trying to cross Manhattan's Columbus Circle on foot; Jesus Christ, sitting through a performance of *Jesus Christ Superstar*—each might be unable to sympathize with our civilization. Schoolteachers prudently withhold from their charges the knowledge of world history, mindful of the danger that present-day life might seem to them abysmally base should they ever learn about humanity's past.

Still, at the risk of disturbing our complacency, it would be worthwhile to expand, or at least to ventilate, the narrow world of our primers, above all, those on architecture and related subjects, and to rouse the student and aspirant builder from the torpor induced by the tedious profundities of declamatory architecture or the numbing banalities of design. He should try to retrace the evolution of architecture with all its bypaths and dead ends down to its very beginnings, either for curiosity's sake or intellectual wanderlust. To add more kinks to his brain's convolutions, he ought, now and then, to forego the study of the accredited architectural masterpieces, old and new, in preference to the unspecific primitive and not-so-primitive buildings of old, and contrast them with our present-day not-so-human ones.

While his mind is still uncalloused, get him accustomed to looking at architecture not as a historical pageant but as an inventory of human shrewdness and stupidity. Give him more options and less opinions. Let him chalk up the gains against the losses, and make a fresh assessment of architecture by leaning less to interpretation and more to contemplation. Instead of Vitruvius give him Pliny so that he may read the tantalizingly terse description of such attractive creatures as the three-span men and pygmies of India who make their houses of mud, feathers, and eggshells. "The world is filled with knowledge; it is almost empty of Understanding," lamented the Chicago sage, architect Louis Sullivan. "For let me tell you, knowledge is of the head, Understanding of the heart."[3]

It pays to swallow these astringent pills of wisdom. The great Puritan experiment to discover the limits of human capacity for taking punishment has not benefited mankind as expected. Anybody who is able to perceive the wretchedness of our way of life and the containers who shape it will want to strive for a more dignified existence. The difficulty lies in finding a way out of our mental slump. Only an inquisitive mind has a chance to rise above the herd. Unhappily, curiosity has a poor reputation among us. Nevertheless, we must cultivate it to keep one step ahead of the animals. Henri Bergson, another man with a philosophical mind, resolved the dominant chord struck by Sullivan into the tonic: "*There are things that intelligence alone is able to seek, but which, by itself, it will never find. These things instinct alone could find; but it will never seek them*"[4] (Bergson's italics).

A note on the illustrations

The documenting of nonpedigreed architecture presents difficulties that are largely absent when dealing with accredited monuments. Only in the past thirty years or so did people other than anthropologists and ethnologists begin to pay attention to the lowly subject. It probably had to wait for our days of compulsory snapshooting.

In pre-Kodak days the traveler would linger over the object of his interest and, if blessed with a spot of talent, commit his impressions, optical and emotional, to paper or to canvas. Although the camera saves him such time-consuming exercises, the photographic image rarely measures up to the inspired drawing. Absorbed in the camera's mechanics, the eye remains unfocused, the mind uninstructed, the heart untouched.

Before the advent of photography, the depiction of commonplace architecture was mostly incidental to religious and genre painting, chronicles of discovery, books of hours, and the like. Occasionally, an ancient bas-relief or a mosaic discloses the shape of a humble house, yet it is events like the happy discovery of prehistoric house models that light up vernacular architecture's remote past. On the whole, though, Western art, especially painting, has been too much oriented toward the grandiose and the theatrical to bother with recording rustic buildings. Even heroic architecture sometimes fared not much better; thus, for lack of authentic portrayals, many a famous edifice—legendary, biblical, or plain antiquarian—exists only as a figment of our mind. Of the Seven Wonders of the ancient world, for example, all but the Pyramids vanished without leaving a clue to their appearance (except the Artemision at Ephesus, a regulation temple on a colossal scale), and one cannot help thinking that, had the Pyramids been razed a couple of thousand years ago, we might not know what they looked like. At any rate, circumspection is in order when examining illustrations made by artists who are not acquainted with the full facts.

In these pages short shrift has been given to the custom of slavishly illustrating the text. Instead, whenever possible, parallels or antitypes are introduced to drive home the unity and constancy of the untutored builders' achievements. Moreover, just as the text is intentionally fragmentary, the illustrations are deliberately promiscuous; any survey of nonpedigreed architecture with claims to completeness would fill the shelves of a medium-size library. Preference has been given to out-of-the-ordinary pictures. Whereas, for instance, the

2. Before the advent of picture postcards, many an image of the world's great monuments turned out to be a fiction of the artist's mind. Illustration from a nineteenth-century Chinese periodical. (Courtesy Columbia East Asian Library)

reader's familiarity with the looks of an English cottage or a Breton chaumière can be taken for granted, sleigh houses may have eluded him. Again, the Swiss peasant houses that appalled D. H. Lawrence need neither description nor depiction, while the amenities of even only mildly exotic dwellings do. For supplementary photographs—some of them rare or previously unpublished—the reader is directed to the pages of *Architecture Without Architects.*

3. Nineteenth-century illustration of troglodytic dwellings in Tunisia.

In praise of caves

Without doubt, the fact that most caves are gratuitously supplied by nature puts them beneath contempt. Snug and delightfully solid as most of them are—if one discounts the occasional rattle of some unrecorded quake—they are not offered for rent or sale, and the idea of inhabiting one never occurs to us. Despite the increasing unhealthiness of our surface life—the dangers compounded of mephitic air and polluted water, not to mention the ever-present dread of atomization—real estate agents so far have overlooked some startling opportunities. They go on peddling the flimsy wooden crate, the plaything of floods and tornadoes, that promises no refuge from an angry Nature. Compared to the rock-bound cave, today's house is as precarious as a canary's perch.

In times past caves served as human shelter much like ordinary houses, although they may not have been regarded as personal possessions. However repugnant the thought of living in the naked cleft of rock may be to us, caves have often been selected by man as a retreat from the intemperance of the weather and as a hiding place from his enemies. It just happens that prejudice is stronger than fear or any practical consideration. To our way of thinking, caves are for cavemen only; troglodytism—living in caves—amounts to disowning one's status as a human being. This calumny is spread by such scholarly publications as the *Oxford English Dictionary,* which defines *troglodyte* as "a degraded person like the prehistoric or savage cave dwellers."

4. Drawings of paleolithic hunters from the caves at Alpera, Spain. (*After Obermaier*)

There is nothing improper about the private world of caves. On the contrary; faith, piousness, and religiosity of all shadings seem to thrive in their padded silence. Prophets, seers, or just any prayerful men and women have always shown a predilection for raw shelter. The unimproved natural cave favors, nay, inspires, mystic communication with rocks, trees, and the sky, the last being but one remove from Heaven. It seems that all hairshirt architecture, from the basic cavern to the treetop lair, with its obbligato of dripping water and bone-chilling drafts, its spiders and bats, vipers and salamanders, is supremely qualified to induce a feeling of light-headedness which furthers meditation. However, a lucid exposition, or even the baldest summary, of the cave's suitability as undesigned shelter is still missing. The subject has never been systematically broken down into species, perhaps because it lacks the spice of human interest, so indispensable by today's conventions for lighting up the past.

The creature whom we label, carelessly, a caveman (a vulgarism that usually stands for upper paleolithic man), was actually an outdoor type, hyperbolically husky, sweaty in an artistic way, utterly normal on the outside yet with demons swarming in his breast. A sportsman with a naturalist's acumen, he led a seminomadic life and probably used caves only occasionally as shelter. To judge from the kind of litter he left behind—mostly meat bones and human skeletons—caves served him chiefly for cookouts and burials. However, unlike us, he did not memorialize the dead with tombstones or painted likenesses. A fastidious and technically accomplished artist, he was mainly preoccupied with the animal world; among the herds of bison and buffalo that he so brilliantly portrayed on the walls of caves, the image of the human hunter amounts to little more than a shorthand symbol. Nor were still lifes or landscapes of interest to him; his painterly vision was Michelangelesque—a preference for the entanglement of supple, albeit brutish, bodies. Indeed, there is nothing facetious about calling Lascaux the Sistine Chapel of prehistory; the famous caves, it has been inferred, were shrines rather than ordinary dwellings.

Still, the stereotype picture that sticks in our mind is mostly the cartoonist's idea of a creature only a few notches above the beasts: a hairy brute swinging a club and dragging his spouse by the hair. Anyone desirous to meet him will have to be satisfied with a glance at his effigy, enshrined in a murkily lit vitrine in some natural history museum. Although his get-up, based on nothing more substantial than the curator's guesses, suggests a low fellow, the facts tell a different story. Not only was he highly gifted, he probably was far more good-natured than today's man; cultural anthropologists have failed to find any signs of human warfare.

5. Monumental bas-reliefs, sculpted by the elements, tower above the entrances to the cave dwellings at Almanzora in Andalusia.

6. Overleaf: The so-called Cliff Palace is one of several hundred cliff dwellings and pueblos in the Mesa Verde National Park in Colorado. Its earliest occupants are believed to have been Mongolian hunters who migrated some twelve thousand years ago. The extensive structures were built by Basket Makers.

Man had to cover much emotional territory before he arrived at the mastery of depicting creatures' shapes and movements. Ages before paintings became portable and suitable for framing, there existed already some sort of fraternal artists' societies. "It was not individual fancy that produced the painted caves, but some institution which directed, and in each period, created uniformity of expression,"[1] suggested the Abbé Breuil, paleontologist and foremost chronicler of cave art; "that there were colleges of artists, far from each other, but subject to the conventions and same fashions, is also certain."[2] The discovery at Les Eyzies of sketchbooks—bone fragments covered with thousands of beginners' "life studies"—would seem to support Breuil's assumption. These artistic finger exercises cannot have but sharpened their observation and probably led to their earliest architectural drawings. Some wall paintings contain numerous tentlike figures, called tectiforms, which are believed to be diagrams of wood constructions. Or they may be rough renderings of huts in which ancestral spirits lived. Besides, a good many of such drawings whose meaning is equally obscure look like fragmentary house plans. Whatever they are, they prove that the virtuosi of figurative painting also drew squares and rectangles, and thus were acquainted with rudimentary geometry.

7. These Magdalenian wall engravings from caves in the Dordogne, France, are thought to represent wooden posts and rafters. They also have been interpreted as diagrams of huts or animal traps, and if they were pits, there may not have been much difference between the two. (After Breuil, Obermaier, and Verner)

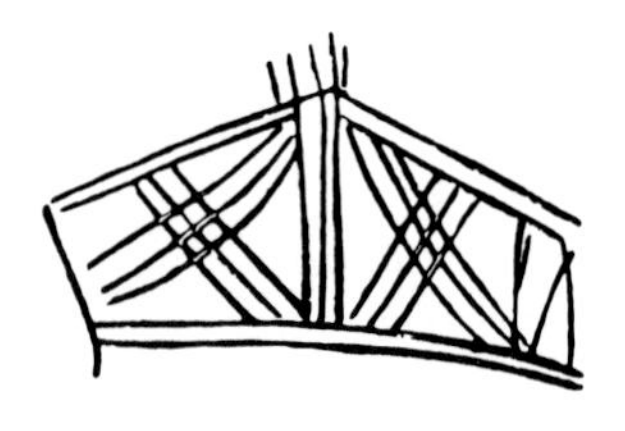

8. Figurative mural by the prehistoric inhabitants of the Canary Islands. (After Hernández Benitez)

It comes as a shock to learn that the brain of the ancient troglodytes of Lozère in southeastern France was found to be larger than that of today's Frenchman. While the cranial capacity of modern man varies between 1,400 and 1,500 cubic centimeters, that of Cro-Magnon man was 1,650 cc and of Chancelade 1,700 cc. Cro-Magnon man, who lived 30,000 years ago, in the opinion of René Dubos "was probably indistinguishable from modern man mentally as well as biologically."[3] Some day science might enable us to peel from the caves' walls layers of sound—small talk, family altercations, chants (of course they did sing)—which would tell us about the inner resources of these people. At any rate, the time has come to dispel the belief that all inhabitants of caves were, or are, subhuman rabble. Quite the contrary; the uncouth cave often attracted an elite such as rarely walked aboveground.

It is a sobering thought that Christ was born in a lowly cave (the Bethlemitic stable as His birthplace is an architectural genteelism),

9. The prehistoric Painted Cave at Gáldar on the island of Gran Canaria is remarkable for its square shape and geometric decoration. This sketch, made in 1873, the year of the cave's discovery, shows a black, red, and ocher mural, since then destroyed by dampness.

10. Rome and Byzantium disagree on the setting of Christ's birth. The thirteenth-century Italian artist who painted this Nativity sided with the Eastern Church by placing the Manger in a cave rather than a stable. (*Courtesy Fogg Art Museum, Harvard University*)

and that His word lived on because His disciples literally went underground. The grotto as a refuge pleasing to God has its precedent in the cave that concealed the infant Zeus, or in the one to which Amaterasu, the Japanese sun goddess, retired, thereby plunging the world into darkness. Porphyrius, commenting on the Cave of the Nymphs in the thirteenth book of the *Odyssey*, notes that caves were consecrated to the gods long before the first temples were built, and Zoroaster dedicated "a spontaneously produced cave, florid, and having fountains, in honor of Mithra, the maker and father of all things."[4] Besides, anyone acquainted with the life and death of ancient Mediterranean nobility, terrestrial or celestial, cannot help being struck by their taste for subterranean residences. Heroes, nymphs, and monsters alike made their home in the folds of the earth. Some of the most famous love matches were consummated in the intimacy of caves: Medea, the lady magician, took Jason in marriage in a cave at Iolcos; Odysseus dallied away seven years on the island of Ogygia as house guest of the cave-dwelling sea nymph Calypso; again, a cave was the classical setting for the union of Thetis and Pelias, of Dido and Aeneas.

Aeneas also figures prominently as guest of another notorious troglodyte, the Cumaean sibyl, Mother Divine of all seers, who preferred the beguiling cachet of a cave to more conventional lodgings. Her sanctuary near Naples consisted of a sort of crypt under the temple of Apollo, presumably built by the aviator-architect Daedalus as a token of thanks on the occasion of his successfully completed flight from Crete. (Aware as he was of the historical importance of this feat, he bequeathed his wings to the holy place, thus adding a touch of the Smithsonian to it.) Although outranked by Delphi, the Vatican of necromantic prophecy, the Cumaean sibyl's cave was a hotbed, if not of intellectualism, at least of divination, smoked and cured over a magic tripod. The words that issued from it seemed as portentous to the ancients as the more expensive oracles of analysts and pollsters seem to us.

There was nothing lowly about the supernatural agency run by the seeress. Its down-to-earth, rock-bottom rusticity was tempered by an aroma of saintly exclusiveness, perhaps because at one point of her career she had been married to Apollo. Visitors to the premises were rattled by the sheer vastness of her cave, laid out as it was on the scale of a concert hall rather than that of a boudoir. According to Virgil, no fewer than one hundred entrances led to the oracle, something that

11. Caves seem to suit Christian saints no less than pagan sibyls. A foremost example of such lack of prejudice is set by the eleventh-century church of Saint-Emilion in the Gironde, France. It is 132 feet long, 66 feet wide, and 53 feet high, hollowed out of solid rock.

taxes the imagination of the most megalomaniac architect.

When the sibyl grew older (she had muffed her chance to ask Apollo for immortality) and her local prestige declined, she packed her things and moved north. This expedient is not unique in the annals of the gods; Venus, alias Aphrodite, similarly prospecting for new spheres of activity, crossed the Alps into Germany where she found a suitable cave in the Hörsel Mountain in Thuringia. However, an alien mental climate and the emergence of an uncompromising foe, the Church, obliged her to adopt novel methods of seduction. The way she handled her most notorious affair, the temptation and conquest of the moody Tannhäuser, bears witness to her unabating resourcefulness. As even the most casual operagoer knows, she supplemented her charms with a *corps de ballet* and, to suit the Teutonic tastes of her lover, a full orchestra of heretofore unheard brassiness of sound and sentiment. Stifled by all this heavy-handed gaiety, Tannhäuser sought a change of air in a voyage to Italy. In Rome he confessed himself to the pope, who promptly refused absolution—whereupon the good knight returned to the enchanted cave to live happily ever after.

Even so, one would be well advised not to take up prolonged residence in Nordic caves. From the point of view of healthfulness, they are no better than dwellings aboveground. Whatever the purported advantages of living in a rude climate—the hardening of body and mind to the point of callousness—cold caves are no place for self-indulgence. For, whereas in a warm climate the temperature of subterranean spaces is nearly constant the year round, northern caves—we only have to think of New York's subway—are equally uncomfortable in summer and winter. *Arthritis deformans,* the bone disease that was epidemic, indeed, endemic, in the Ice Age, afflicting cave man and cave beast alike, is still with us.

On second thought, the caveman's cave may have been less uncouth than we imagine. It even contained something like a private shrine, a niche for the "Venus figure," comparable to the *Herrgottswinkel,* the Lord's corner of the alpine peasant, or the Buddhist shrine of the Japanese. Or, for lack of it, the hearth doubled as altar. With but a rudimentary kitchen at hand, our troglodyte may have chanced upon eating his meat *saignant,* his vegetables intuitively undercooked, all without benefit of a nutritionist's advice. Alas, no scholar has set himself the task of reconstructing, if ever so sketchily, a conspectus of Stone Age cooking. No menus are extant from prehistoric times, yet the daily fare of a cave family remains unsurpassed. ("Who among us ever tasted the delicacy of delicacies, boar's snout?" I asked once. "Surely, the culinary athletes of bygone days would have shuddered had they been offered the pap that comes in our cardboard

12. Seminaturalistic Iberian cave painting of stag, hind, and fawn. (After Breuil)

containers and tin cans. We shall never recapture the fragrance of the cavemen's dishes; we must content ourselves with feasting our eyes on those supermarkets in effigy; boars and oxen, red and fallow deer—a Protean protein diet. . . ."[5])

The fires built in the cave were probably no more effective in dispersing cold and damp than our electric heaters. What kept them burning was the need for smoke rather than warmth. Not only did smoke act as a disinfectant, it preserved the stores of Stone Age larders: dried fruit, dried fish, and dried meat. It cleared the cave of vermin and of all animals with a lower coughing point than man. On the whole, though, troglodytic fauna held few terrors. In time the dragons and minotaurs departed from the scene, while scorpions, vipers, and poisonous spiders, despite their alarming reputation, were never much to worry about. What ultimately marked the domestication of the cave was the appearance of domestic animals: a sweet smell of cow and donkey, similar to that of the Manger (before the Magi spiked it with frankincense and myrrh) pervaded the premises.

Besides serving as *pied-à-terre* for philandering gods, caves also accommodated a galaxy of demigods and their demimonde, as well as some of the great of the earth. In times when emperors were more numerous than they are today, tradition prescribed that they enhance their image with expensive building programs, when actually their private demands were on the frugal side and all they craved was a comfortable cave. Emperor Tiberius, to name but one, was an aficionado of underground apartments. He shared this taste with the Roman landed gentry, who placed their country seats in the vicinity of a *specus aestivus*, a grotto that promised coolness during the most torrid summer. Unfortunately, these elegant cellars had one annoying drawback: Italy's volcanic soil is prone to the ravages of earthquakes. Tacitus tells of a dinner party in Tiberius' cave near today's Sperlonga (the name is derived from the Latin *spelunca*, cave), when suddenly the roof gave way, burying a number of guests and attendants. The monarch himself remained unscathed, thanks to Sejanus, his chief minister, who had the presence of mind to throw himself over Tiberius, protecting him with his body.

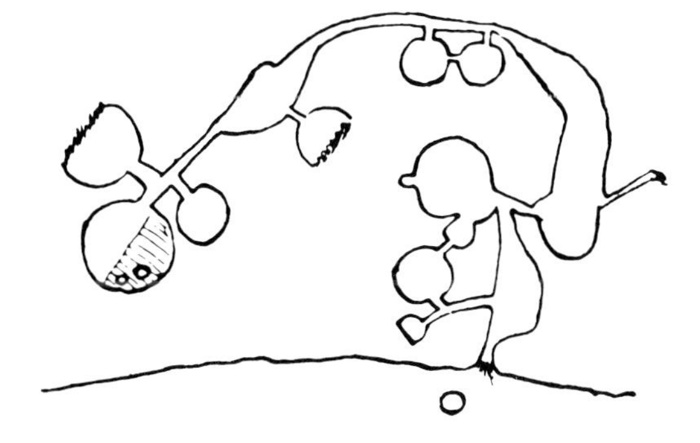

13. *Plan of a cave close to the ruins of a Roman villa at Cluseau de Fauroux in France. The circular rooms are man-made and in summer may have served as a cool annex.* (*After Blanchet*)

A subspecies of fastidious cave dwellers are those sovereigns that took up residence in caves *after* their demise. This accords with the popular belief that immortal men live in sumptuous surroundings deep underground. The Holy Roman Emperor Frederick I, alias Barbarossa, is sleeping under the Kyffhäuser Mountain in Thuringia, waiting to be awakened to unite the German peoples. A variant of this legend makes Frederick II the tenant of the same mountain; national heroes are apparently interchangeable. An earlier epic emperor, Charlemagne, inhabits a hill in Westphalia, as does his adver-

14. Although these Aegean habitations were built and carved without level, plumb line, rule, or compass—the curves are those of a loaf of bread rather than geometric ones—the high sculptural quality of walls, vaults, and staircase belies the supposed primitiveness of all cave dwellers.

sary, the Saxon Widukind. Henry I is hibernating in a cave near Goslar—and so forth. It all goes to show that underground lodgings carry a prestigious address.

Hermits of every philosophical persuasion were another group with a distinct hankering for subterranean, or at least cavernous, quarters. Whatever their capacity for the self-denial of life's amenities, some kind of shelter was indispensable to their survival. Yet building a house, or even only a roof, they felt, might impair their unworldliness. Hence a natural cave—sometimes improved beyond recognition—afforded them enough protection from the elements without compromising their vow of poverty. It satisfied their idea of unpropertied existence; they remained Nature's squatters.

The picture of the hermit in front of his rocky lair, painted a thousandfold in the art of the Church, may not appeal to us. To our mind solitude fosters indolence and vice; relentless meditation seems nothing short of consorting with the devil. Worse, the recluse shirks competition, to us the foremost aim in life. Be that as it may, some of these petty-troglodytes, confident that their instincts were sound, advertised in no uncertain terms their cave's coziness (see page 81). The geographer Agartharchides, who lived in the second century B.C., filled fifty pages with descriptions of cave dwellers in the Red Sea region alone. Moreover, Palestine, Anatolia, Ethiopia, and the countries of the Mediterranean all the way to the Pillars of Hercules were rich in natural caves. One modern geologist even speaks of troglodyte *nations*.

To be sure, these were no nations of Calibans. Not all cave dwellers affected the scanty dress of early man; they were probably indistinguishable from ordinary people. Their diet was—at least to our mind and palate—on the adventurous side. Ethiopian troglodytes, for instance, fed on serpents and lizards. Others drew strength from drinking a mixture of milk and blood. But let nobody tell you that the prophets in the desert underwent culinary hardships simply because, as the Bible notes, they sometimes had to subsist on locusts. Not only is the taste of toasted grasshoppers exquisite, but one may gorge on them while appearing to undergo physical mortification.

To dispose of another popular misconception—cave-dwelling hermits are not necessarily misanthropes. Although conviviality would seem to defeat the very idea of self-imposed solitary confinement, history shows that reclusion is quite compatible with togetherness. The best example, a veritable apotheosis of troglodytism, is furnished by a former colony of monks and nuns in Cappadocia, a wild mountainous district in today's Turkey: To the west of Erciyas Daği, the Mount Argaeus of the ancients, lies a volcanic region that is among the most attractive in the Old World. The soil of the valleys supports

orchards and vineyards of great fertility, while the conical hills are riddled with cavities eminently suited for human dwellings. The fame of the place rests however on a gigantic housing project put up by Nature at her most magnanimous. But let me yield the word to one Paul Lucas, an emissary of Louis XIV, who claimed to have been the first European to lay eyes on the extraordinary sight. "I was overtaken," writes Lucas, "by an incredible astonishment at the sight of some ancient monuments on the other side of the river [Kizil Irmak, or Red River, the Halys of the Greeks]. Not even now can I think of them without being struck. Although I have traveled a good deal, I never have seen or heard of anything similar—pyramids in prodigious number and of various height, each consisting of a single rock, hollowed out in a way as to provide several apartments, one on top of the other, with beautiful entrance doors and large windows for the well-lighted rooms."[6]

15. Wind and rain have shaved off the outer wall of this triplex apartment.

Had Lucas fallen victim to an optical illusion, or did he merely suffer from the kind of travel euphoria that sometimes befogs the mind of the most sober observer? He had not; the pyramids were palpably real, although somewhat less perfectly shaped than Lucas's drawings would have us believe. For all we know, he may very well have made out doors and windows where others saw but gaping holes, if only because he considered doors and windows the proper attributes of all grand architecture. Having taken a fancy to these *maisons pyramidales*, he mentally elaborated on their design. "I mused at length about the meaning and purpose of so many pyramids, for it was not just a matter of two or three hundred but of more than two thousand. [Later, he raised his estimate to twenty thousand, which is still short of the actual number.] Each of them terminated in a cowl or bonnet, shaped like those of Greek priests, or a female figure cradling a child in her arms, that I immediately took to be an image of the Virgin. Which gave me the idea that they once sheltered some hermits."[7]

16. Tufa cones, topped with the sort of lava boulders that Paul Lucas mistook for religious statuary.

17. Above: The 1712 engraving from Lucas's Voyage dans l'Asie Mineure *presents a highly streamlined version of the Cappadocian rock dwellings. Yet, as the photograph in figure 15 shows, his description and graphic rendering are not far from the truth.*

18. Overleaf: The streetless villages in the region of Ürgüp in Cappadocia consist mostly of inhabited tufaceous cones. A sprinkling of free-standing houses adds to the visual confusion.

Alas, for all his perspicacity, Lucas was much akin to the hurried tourist of our day; he simply did not take the time to examine his discoveries. Darting to right and left as fast as the pace of his caravan permitted, he caught a few glimpses—through open doors—of mural paintings and archaic inscriptions, before, to his "mortal regret," an approaching band of highwaymen forced him to abandon his exploration. Little did he suspect that the caves contained acres of murals that some day would open a new chapter on Byzantine art.

At home Lucas's account of the Cappadocian cave dwellings was received with disbelief and his renderings ridiculed. In his excitement he had quite pardonably exaggerated some aspects of the scene—what seemed to him sculptured madonnas were simply lava blocks resting on top of the tufa cones, a well-known geological phenomenon—but his general description was essentially correct. If anything, it fell short of the fantastic reality.

19. Opposite page. This tower apartment, one among thousands of similar ones in the volcanic region of Cappadocia, resulted from the cooperation between nature and man. The exterior shape and substance were gratuitously furnished by the elements, while an assiduous carver supplied himself with living quarters on several levels.

20. Ruined frescoes in one of the three hundred sixty-five rock-cut churches.

Taking a close look at what are now but empty shells of a once teeming warren, one still finds much cause for wonder. Some of the earth cones (but only a few) seem so perfect as to have been turned out on a potter's wheel. In a way they look like petrified tepees, except that they sometimes reach the height of a fifteen-story building. The shapes are Nature-made, the result of erosion by wind and water. Cracks and holes dot the surfaces, suggesting some elementary shelter. But perhaps even more than these, the finished silhouette of each peak spells out a *house*. To a man looking for a roof over his head, the rocks were a godsend; he was easily tempted to deepen the natural recesses and cut out, so to say, a niche for himself. Since the

21. *A cross-section, as precise as a draftsman's drawing, reveals Cappadocian cave dwellings in an advanced state of disrepair.*

22. The caves that line the southern border of the Indre in the Touraine, France, are used for wine cellars as well as for dwellings. Figures 23 and 24 show them at close range.

porous rock is no more difficult to cut than hardened cheese, a man can in no time and without much exertion excavate a good-size apartment, including some stony furniture: tables, benches, and couches, not to mention fireplaces.

The fascination with caves has not worn off in our days. The current interest in spelunking, oriented as it is toward the decorative, such as stalactitic caverns, parallels the enthusiasm for mountain climbing that swept the world in the eighteenth century, when breaking his neck in the pursuit of alpinism enhanced a man's reputation in much the same way as being killed in a duel. Today, when mountains have lost their terror and mountain climbing has turned into a sport,

23, 24. Close inspection of the houses shown in figures 23 (opposite page) and 24 (below) reveals hybrids of trianons and cave dwellings. Naked rock blends into polished façades with dainty moldings, portes-fenêtres, and curved flights of stairs, while an impenetrable thicket substitutes for a conventional roof. The result is surrealistic architecture, reassuringly cozy.

25, 26. Contemporary architects are not adverse to borrowing ideas from troglodytes. The dwellings of this Chinese rural community are dug into soft loess and grouped around 30-foot-deep courtyards, accessible through L-shaped staircases. The UNESCO headquarters in Paris (opposite page), similarly tucked away below ground, have the unique distinction of remaining unsuspected by, indeed, invisible to, passers-by.

cragsmanship takes to tenebrous zones; the trend is from the Olympian to the spelean. Yet whereas in warm and temperate zones caves have always been valued for their usefulness, the purpose of spelunking now often is no more than an endurance test.

Still and all, and considering that in a country with the appropriate soil it may be cheaper to carve a house out of the ground than to build one—no longer with a whittling knife but perhaps with the help of an attachment to the lawn mower or vacuum cleaner—the question of whether caves ought to be earmarked as general habitations is not as absurd as it may sound. Certainly, the theory that caves might serve as fallout shelters has long been discredited. Nevertheless, if we can read forebodings, a far more awesome prospect than incineration is in store for humanity. To all appearances the time is not far off when our planet will cease to provide as much as standing room for its inhabitants. Even if people were able to adapt themselves

to a double- or triple-deck civilization, there is a limit to the rank growth of skyscrapers and multilevel highways. The only way out of the human rabbit warren is, quite simply, down the rabbit hole.

Automobiles set the trend long ago. Too bulky and too numerous to find adequate room under the sky, they have taken to tubes and subterranean garages. We may not want to join them quite as yet. Still, whatever the plans for repairing the damage inflicted on the land, chances are that the native genius for destruction will not be daunted by pious vows and mammoth budgets. Free enterprise will prevail. Eventually, though, with nothing left for us to destroy aboveground, we will reassert our proverbial pioneering spirit and depart for the virgin territories of a yet unexcavated netherworld.

The mode of life of all the human and superhuman moles in space and time—the anchorites, gypsies, and trolls of legend—would hardly justify an excursion into the realm of caves were it not that, unsuspected and unknown to us, today many million people are housed underground. So ardently do we cultivate an ignorance of the ways of life in countries whose inhabitants do not share our philosophy, or the lack of it, that few of us have ever heard of the existence of subterranean *cities*. They bear no resemblance to our world of bargain basements and subway stations but have been planned from scratch as modern metropolises, complete with government offices, factories, schools, hotels, and habitations. Over the past forty years such cities have been built in northern China. (The term *northern* is only relative, since the latitude corresponds to that of the Mediterranean.) In the provinces of Shensi, Shansi, Kansu, and Honan more than 10 million people, or about the combined population of Austria and Switzerland, live technically as troglodytes although their standards of comfort and hygiene may not differ much from ours.

This architecture which I shall call deep-dig—in contrast to high-rise—has an honorable tradition. Its beginnings are lost in the mists of history. The *Book of Songs*, the *Shi-king*, mentions one Duke T'anfu who "baked cave dwellings." Which probably means that these caves were then much the same as they are today—cavities in the rock, natural or man-made, lined with sun-dried brick. "The atmosphere is pleasantly cool, and one has the feeling of staying in a regular, massive building,"[8] observed an old China hand, the geologist Ferdinand von Richthofen. However, like massive buildings, cave dwellings are apt to collapse in earthquakes. In the Kansu quake of 1920, 246,000 people perished.[9]

The earth's convulsions never deterred the tillers of the soil. Just as the peasant who lives on the slopes of Mount Vesuvius returns to his land after every one of the volcano's eruptions, the Chinese stick to their perilous caves. "When one looks at this immensely fertile land," wrote Richthofen, who toured China more than a hundred years ago when it was still hospitable to foreigners, "one asks oneself where the people live who cultivate the fields. There are no houses in the valleys nor in the walls of rocks that confine them. The puzzle is solved on approaching the cliffs which buzz with life like a hornet's nest. Innumerable openings lead to ample apartments with windows and doors lined in masonry. In front of the doors are walled courts with threshing floors and forage for livestock."[10]

Northern China is loess country. True loess, which ranks lowest on the geological scale of hardness, is an aeolian or wind-borne rock, with physical properties similar to those of soft volcanic earth. Rivers saw deeply into its surface, and so do roads by the erosive action of wheels. Roads form the narrowest of canyons, many stories deep and hidden from view. The villages are equally invisible because farmers are unwilling to waste land by putting houses on it. To have their cake and eat it, they make their homes in the bowels of the earth. Staircases and sunken courts for admitting daylight and air are often the only link with the upper world. "One may see smoke curling up from the fields," writes the geographer Cressy; "such land does double duty, with dwellings below and fields upstairs."[11]

A splendid concept, but it seems to have no application for our country. For all we know, agriculture is on the way out. Steadily declining, another victim of progress, it may survive as a hobby at best. But the land gained thereby will come in handy. With all architecture safely tucked away below ground, we should be one step nearer to solving the seemingly unsolvable situation of surface traffic. When the entire land will have been asphalted over, making any road system obsolete, cars can move freely in all directions.

As yet we may not be ready for troglodytism. But sooner or later it will become the big issue of national, indeed global, defense—not against the onslaught of foreign nations but against that foe in our midst, the implacable automobile. Our survival will be determined not by abstract ideologies but by the race between the stick-in-the-loess and the builder of castles in the air.

27. *An earthbound* déjeuner sur l'herbe *is but a poor substitute for a picnic in the lofty sphere of trees. The Chinese in this picture, fastidious eaters that they are, may find their appetite stimulated by ever-so-subliminal memories of a former arboreal life. From the nineteenth-century Chinese periodical* Tien-shih chai hua-pao. *Government reprint, 1910. (Courtesy Columbia East Asian Library)*

Brute architecture

In the 1920s, during that uneasy breathing spell between the two world wars which changed forever the complexion of Western civilization, most of life's pleasures were still easy to come by. In Paris, then the hub of the world and fountainhead of *savoir-vivre*, the pace was set not by machines but by man. Work seemed tolerable, and leisure was far from mechanized. A middle-class family or a pair of lovers, desirous to enjoy a glorious day, would go for an outing to what were then, and to some extent still are, the city's bucolic environs. Moreover, they would go by streetcar, that dependable, economic, and sociable means of transportation, now moribund or defunct. Only a few miles beyond the outer boulevards the countryside is thick with châteaux and their concomitant gardens, parks, and forests where strollers liked to lose themselves. Besides, there were any number of small towns that could easily absorb a holiday crowd. One of the more unusual of these picturesque places was Robinson, eight miles outside of Paris, where people went to dine *in trees*.

The first arboreal tavern auspiciously opened in 1848. It was called "À Robinson," probably alluding to the immortal story of urban man facing nature's fitful blessings, and such was its success that others followed suit and the village itself assumed the name. Robinson's airy restaurants held no unpleasant surprises thanks to a benevolent climate and honest cuisine and that noble breed, French waiters, who would serve an eight-course meal, then de rigueur, on a limb of an enormous chestnut tree without spilling as much as a drop of soup.

N XXXIII.
ad pag: 856.
1
Bewohnte Bäume in America. I.
2.
II.

"After having themselves restored between heaven and earth," recounts a French guidebook, "after a dance outdoors, a game of *boule* or *tonneau*, or a go at the swing, the cheery couples would set out into the countryside on mules or horses."[1]

Today, when mass murder on the highways has become the accepted Sunday observance, the amusements of another day seem disconcertingly quaint and not a little suspicious. Robinson's exotic restaurants are still a going concern but, as the above-mentioned guide notes with regret, the attendant pleasures have been industrialized. Yet the fact that partaking of a meal, squirrellike, in a tree, appealed to seven generations of Parisians, is so foreign to our ways that it calls for an explanation.

On the whole, the French have never been known to be eccentrics. If anything, they tend to be sticklers for correctness, and nothing could be further from their mind than to demean such serious business as dining on good food by mixing it with fun and farce. To be sure, anybody with his senses intact prefers dining in the open air to dining in a smoky, smelly room, but then a classical *déjeuner sur l'herbe*, alias picnic, ought to meet his needs. Indeed, picnicking, whether peasant style or, more formally, on rugs and cushions, or, perish the thought, on collapsible furniture, has not altogether lost out to the attractions of drive-in roadside eateries. Nevertheless, a craving for treetop dining is unusual, not to say seditious.

To be sure, tree-dwelling individuals and societies have existed all along; trees form a natural roof. Besides, animals lived in trees long before man did and their example cannot have escaped him. The Welsh bard and wizard Merlin, hero of Arthurian romances, lived with a wolf for his companion in the trunk of an old oak. Similarly, the African baobab tree, which attains a diameter of 30 feet, provides a spacious shelter when hollowed out. Such lend-and-lease from Nature never appealed to southern nations. "The Fenni [today's Lapps] live in astonishing barbarism and disgusting misery," wrote Tacitus in the *Germania*; "they have no fixed houses, nor have their infants any shelter against wild beasts and rain, except the covering afforded by a few intertwined branches."[2] The Romans of old may have mouthed bucolic Greek lyrics, but they liked their houses solid.

Li-chi, the Chinese ritualistic book, states that in remotest antiquity, that is to say, before people knew how to make fires, "the ancient kings had no houses. In winter they lived in caves which they had excavated, and in summer in nests which they had framed." The trees they inhabited were their green mansions and vicarious palaces. So they were to the Tasmanians, now extinct, whom James Cook found living in trees "like fauns and satyrs." And before Cook, Sir Walter Raleigh reported of a "very goodlie people," the Tinitinas,

28. Opposite page: Some indigenous Americans nested in trees much like birds or apes, a custom that cannot have but annoyed the European invaders. The two engravings, captioned "Inhabited trees in America," are taken from Erasmus Francisci's Lustgarten *(1668). (Courtesy Rare Book Division, New York Public Library)*

who "dwell upon the trees, where they build very artificiall townes and villages."[3]

It is one thing to ruminate about sylvan spirits, it is another to convince ourselves that anything as prickly as the woodlands contribute substantially to the comforts of body and soul. Yet there was a time when people looked with wonderment at a young shoot; when foliage was solemnly strewn into a newborn's cradle as a stylized gesture of nest building, or as an allusion to an outdoor temple's coppice.

Our maypole celebration is a survival of tree worship that paralleled stone cults since earliest times. Savage and civilized societies alike revered trees and groves, some of which were permanently or temporarily inhabited by gods and godlike creatures. The first temple, scholars tell us, was a tree; a god's image was either enclosed in a hollow tree or sheltered under a tree's roof. Even after the departure of the gods this usage did not die out altogether; in fact, trees are still honored by courtesy of the Church. In the Catholic countries of

29. The image of the Virgin conveys an odor of sanctity to the willow tree, which repays the compliment by forming a live halo around it. From Arte Popolare Italiana *by Paolo Toschi and Carlo Bestetti.*

30. Tree worship is condoned by the Church if combined with the adoration of holy pictures. The Pino Santo, the Holy Pine on the Canary island of Palma, sets an example of this judicious compromise.

31. Below: By dint of architectural acrobatics a dead tree in a Chinese village has been turned into a vertical holy precinct. It supports three solidly built little shrines and a fourth one located inside. From the Nouveau Journal des Voyages *(1882).*

Europe and South America, where official religion has not completely stifled nature worship, people strike, if unconsciously, a compromise of loyalties by affixing the pictures of God and saints to trees. The honey of godliness, a transferable substance, thus flows like sap in the chosen tree.

For an early example of tree worship and tree dwelling: at Dodona in Epirus, the seat of the most ancient of Greek oracles, Zeus roomed in the trunk of an old oak, much like an African native may live in a hollowed-out tree. The holy oak also doubled as Zeus' spokesman. Although it did not possess the power of speech, it rustled meaningfully when asked a question. However, the Greeks had no monopoly on arboreal intelligence; in Arabia, trees dispensed medical advice to the sick who slept under them.

The Patriarchs, no less than their pagan contemporaries, worshipped in holy woods, and Zeus' oak had its counterpart in the oak grove of Mamre where God received Abraham (Gen. 18). Hence the awe in which trees were held by the truly pious. To break off a twig from a holy tree was ill advised, for wanton damage to trees called for divine retribution. Even in our days there are still woodsmen who pray and ask forgiveness of the tree they are about to fell.

The sacred grove was by no means a substitute for a temple. The wood *was the temple,* its trees the columns and the firmament its roof. The word *templum* signifies a section, a district, a visual field on earth or in the sky; by extension, a piece of land dedicated to a

32. *Ancient Prussian* Rundling, *a circular village with sanctuary, idols, and holy fire in the center.* (*After Hartknoch*)

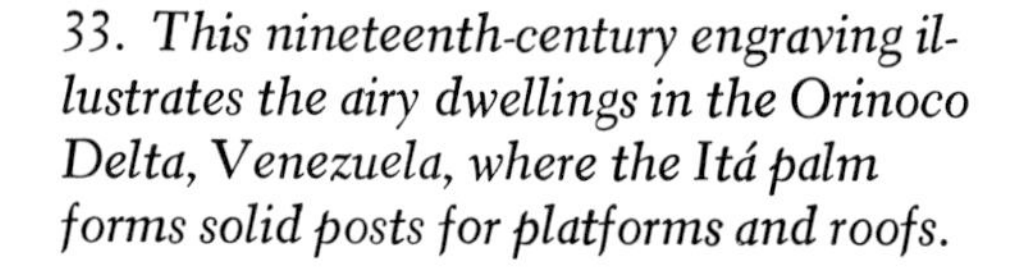

33. *This nineteenth-century engraving illustrates the airy dwellings in the Orinoco Delta,* Venezuela, *where the Itá palm forms solid posts for platforms and roofs.*

godhead, a holy precinct. Most houses of God, whether temples, mosques, or churches, betray their vegetable origin by being orientated and opening up, like a sunflower, to the sun.

Thus what we call a temple is actually the abstraction of a grove; the thicket of columns recalls the thicket of trees. (Some Greek temples luxuriated in more than a hundred columns.) Unlike the oak of Dodona, however, marble columns neither rustle nor impart revelations to people who sleep near them. The leafy sensuousness has been sacrificed to stony precision, the mythical feeling of space to mathematical expanse. Not surprisingly, temple columns ended up as the tritest of architectural clichés.

Quite apart from their symbolical connotations, trees are among the most inviting, not to say poetic, of ready-made domiciles. To dwell aloft in an arboreal rigging and gently sway in a breeze, with the sunlight percolating through a leafy canopy, appeals to an infinite number of creatures, including some of the less inhibited kind of man.

Family-size treehouses are at home in the steamy tropics of Africa and Oceania. Their remoteness—some of them perch fully 80 feet above the ground—is due less to their tenants' desire to obtain a detached bird's-eye view of the world below than the need for protection from marauding animals and hostile neighbors. Insects, hovering low, also are liable to drive a man into a tree. In the Orinoco delta, for instance, human life is made miserable day and night by clouds of mosquitoes. Since, however, they lack the endurance for long flights over water, the natives make their dwellings—mere platforms, with hammocks their only furniture—in flooded palm groves, the live trees serving them as posts.

The urbanite, for want of Tarzan's buoyancy, has long forsaken bucolic dreams. Not so the child of the pseudoarboreal genus. With his bent for leprechaunish adventure unimpaired by years in nursery and kindergarten, he craves a treehouse that he can enjoy, animallike, in secrecy. When his plea for a hideout falls on deaf ears, he is known to become miraculously self-sufficient and to build, if not a house, a shaky lair in the branches of some backyard tree to which he likes to retire with a sweet or a morsel of food. For all we know, it is still the ape in him that prompts him to climb a tree. Or perhaps he shares a dryad's hankering for a sylvan retreat. Is it too farfetched to attribute

34. Tree dwellers on the Orinoco River. From Hulsius's Fünffte kurtze wunderbare Beschreibung . . . *(1603).*

the treehouse's appeal to some atavistic instinct? Could it be that, unknown to ourselves, we haven't quite lost the memory of our erstwhile arboreal habitat?

At some crossroads in prehistory, when the human tree dwellers opted for coming down to earth, the anthropoid apes stayed aloft. Like the Orinoco people, they sleep in trees where they make a platform, never less than 20 feet above ground, and occasionally three or four times that high. A particularly comfort-loving fellow will go to the trouble of bending a tree's saplings to form a roof overhead, but on the whole, ape architecture remains elementary. The exception is the nocturnal aye-aye. This peculiarly socialized lemur, A. H. Schultz (*The Life of Primates*) tells us, "builds itself a quite elaborate, globular nest from turds it cuts and carries to a fork high up in some tall tree. The nests are completely roofed over, have a small side entrance and a floor lined with shredded leaves, and provide room for only one adult."[4]

Although apes seem to be well satisfied with their simple quarters, they permit themselves a luxury undreamt of by our most sanguine advertising men. The chimpanzee, a confirmed hedonist, never sleeps twice in the same bed; the least exacting bachelor makes it a point to daily construct a new one. Like those fashionable ladies of yesteryear who would have died rather than be seen twice in the same dress, he cannot bring himself to revisit his bed of a night. In other words, he embraced the tenets of perpetual change and conspicuous waste long before industrial man took them up as gospel.

35. *Nest dwellers on a South Sea island. From* Tien-shih chai hua-pao. (*Courtesy Columbia East Asian Library*)

The apes' nocturnal extravaganza is remarkable enough, but what is one to make of the fact that they also build a separate day nest! Obviously, some apes have solved problems of leisure that our conscience forbids us to even only contemplate. As befits a breed with a philosophical bent of mind, at some time between 9:30 A.M. and 3:00 P.M., apes spend two hours daily for a siesta.[5] It is an institution of long standing, and although it equally benefits man, it is odious to people to whom time is money. Despising Latins as we do for taking a nap after lunch, we might be more inclined to aping the apes. With the number of working hours dwindling to ever new lows, and machined entertainment becoming a deadly bore, we may yet have to accept the siesta as a means for making a day fit into twenty-four hours.

When we say we make our bed, it is strictly a manner of speaking. The ape on the other hand actually does make his bed from scratch every day and has never been known to tire of this self-imposed task. He is so good at it that it takes him no more than five minutes. Moreover, he always succeeds the first time. Or almost; only rarely has he been seen to throw up his hands in despair at some unwieldy branches. A born builder, he does not need much parental guidance. "In the wild," observed Irven DeVore (*Primate Behavior*), "the infant has opportunity for watching nest-making, and infants still sleeping with their mothers sometimes make little nests during the day as a form of play activity."[6] Even the captive ape, though separated from his mother at birth and far removed from his native habitat, is seized by the urge to make his nest regardless of an uncongenial environment and inappropriate building materials. Marooned on a hostile floor, with but a few sticks and stones at his disposal, he plots an *abstract* shelter. He uses his scant supply of material for creating a ring around himself and, if he runs out of material, "the ring is the only thing that is made. The chimpanzee," wrote Wolfgang Koehler, a dedicated ape watcher, "then sits contentedly in his meager circle, without touching it at all, and if one did not know that this was a rudimentary nest, one might think that the animal was forming a geometric pattern for its own sake."[7] Avebury and the like are probably man's answer to the chimpanzee's ring.

Alas, the apes' productive span is short; as they get older, their creativeness dries up. Whether this is a retirement syndrome or premature discernment of art's futility, nobody knows. Still, we had better be informed about the finer points of animal architecture and engineering, not to do as the animals do, but in order to preserve our humaneness. On the whole we can learn from animals a good deal more than they can from us. Whereas man is unable to shape a tool or build a house without previous experience, most animals have an innate sense of construction. As Darwin noted, a beaver can make a dam, a bird its nest, a spider its web at the first try. Although the web and the average nest are ephemeral, the dam is not. Beavers' consolidated bulk of work will stand up to the ravages of time for thousands of years, a record rarely matched by man-made constructions, perhaps because beavers are eager to provide constant supervision and labor for repairs, while man is not. Equally remarkable is the fact that the beaver's dams are by no means a vital necessity. All he needs are natural ponds or river banks for his burrows. Hence, as one professional admirer of the American beaver pointed out, what lends him an almost human quality is "that he should have voluntarily transferred himself, by means of dams and ponds of his own construction, from a natural to an artificial mode of life."[8] In other words, the beaver betrays a downright human trait in his zeal for building in excess of his needs.

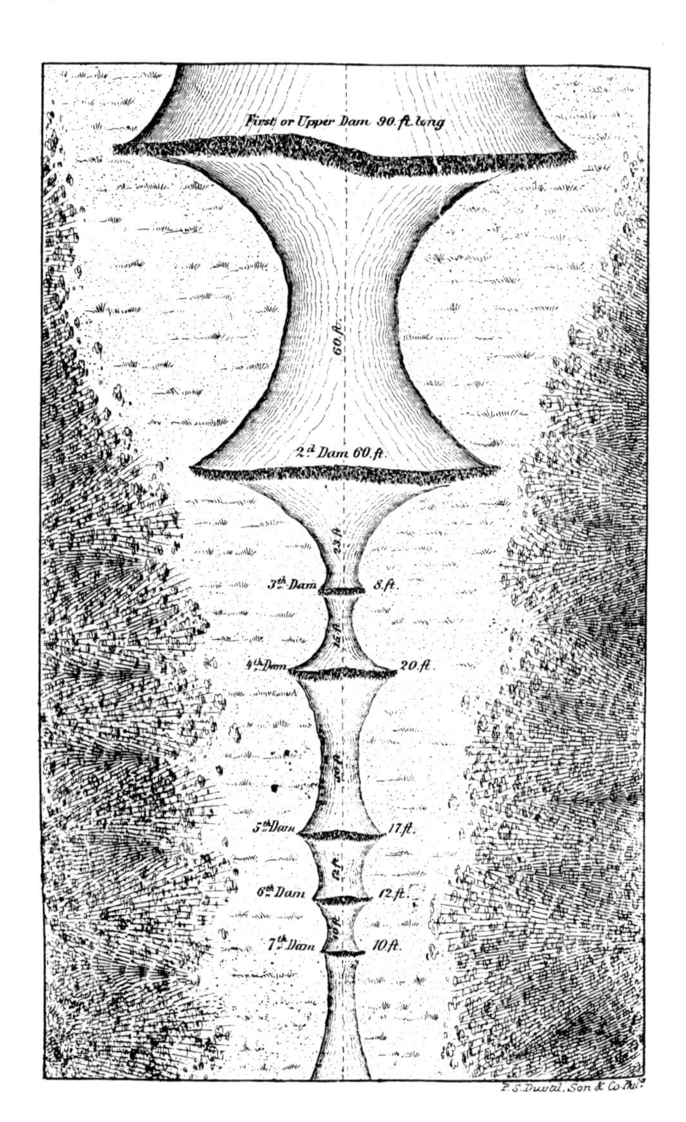

36. *A sequence of seven beaver dams at the entrance of a gorge, the largest of which forms a pond covering 10 acres. From* The American Beaver and His Works *by L. H. Morgan.*

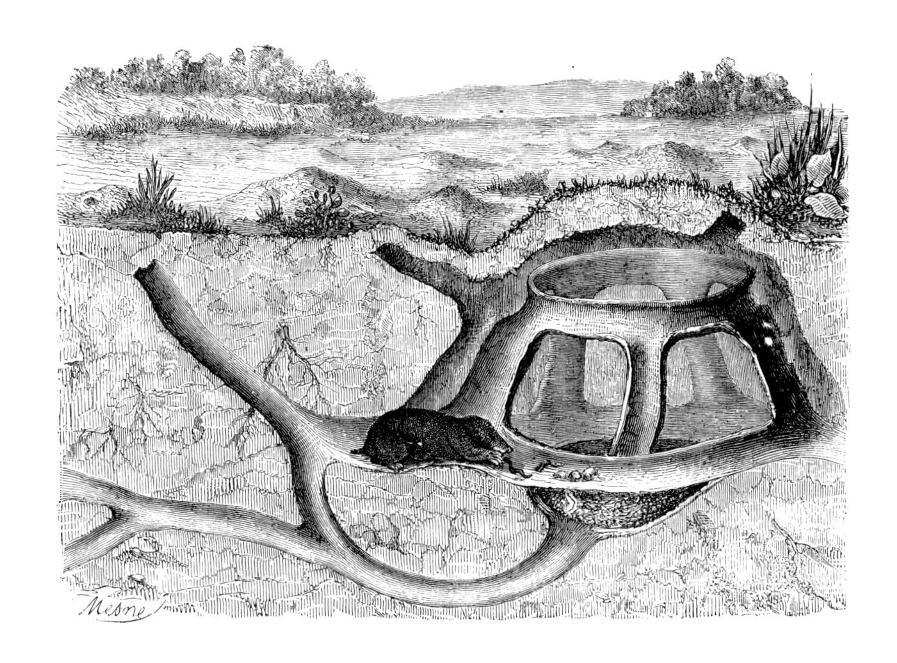

37. *Moles do not just dig tunnels and wells, they construct stylish underground dwellings.*

Many beasts have an uncanny talent for engineering; not only did they invent all sorts of tricky walls and roofs, they sometimes skipped —as did the termites—our cumbersome traditional techniques and went right on to mastering highly sophisticated modern building methods. Unlike man, most animals have a definite conception of what makes a perfect shelter. Conceivably, they just use their brain. "No truth appears to me more evident," Hume declared, "than that beasts are endowed with thought and reason." Since natural instincts are lost under domestication, man, that most overdomesticated mammal, has nothing to fall back upon but a blend of wishful thinking, of luck, good and bad, and what he calls his experience. His brain is poorly aided by his five senses. His nose, having exhausted its utility, is mainly ornamental. He cannot hear the dog whistle he blows. He lacks the sonar system that guides the bat in the dark or the visual acuity that permits the bird of prey to spot its victim. He lacks the animals' inherent know-how.

Granted, animal builders work under enviable conditions. Unhampered by red tape, and innocent of profit motives, with an incalculable backlog of practice at their disposal, they often attain perfection by simply following their instinct. As a rule, their habitations are well suited to their needs, indeed, so much so as to be unimprovable. The shells of certain mollusks for instance, surpass in conception any of man's precision-tooled constructions. People who had the advantage of an old-fashioned education with its teachings of transcendental

geometry and mathematics stand in awe before the houses of the ammonites, that is, not of the biblical tribe but that branch of Cephalopoda which became extinct at the end of the Cretaceous period. Marvelous to relate, the ammonites built their shell around the axis of a logarithmic spiral. They did not wait for John Napier (1550–1617) to invent logarithms but were conversant with them long ago. For all we know, they even may have had a computer-type device integrated into their gut. Molecular structure, equations and theorems, the relationship between stress and strain—these seem to have been to them as transparent as a prepollution drop of water.

Another example of advanced design is the honeycomb. "He must be a dull man," wrote Darwin in the *Origin of Species*, "who can examine the exquisite structure of a comb, so beautifully adapted to its end, without enthusiastic admiration. We hear from mathematicians that bees have practically solved a recondite problem, and have made their cells of the proper shape to hold the greatest possible amount of honey, with the least possible consumption of precious wax in their construction."[9] Would to God that our architects had that much bee sense and obeyed similar principles of economy!

38. *Opposite page: Like many other types of Old World architecture, this Anatolian beehive is closely related to funeral monuments; its shape echoes Lycian pillared tombs. The 14-foot-high pedestal puts the honeycombs beyond the reach of human and animal looters. (The platform is reached through a shaft within the pillar.) Hollow tree trunks, covered with wooden shingles and bark, shelter the bees. Several such apicultural structures stand in the pastures near Elmali (see p. 189), not far from the Mediterranean shores of southwestern Turkey. From "Lykische Gräber; ein vorläufiger Bericht" by Jan Zahle,* Archäologischer Anzeiger *(Berlin), 1975. (Courtesy Professor Machteld J. Mellink)*

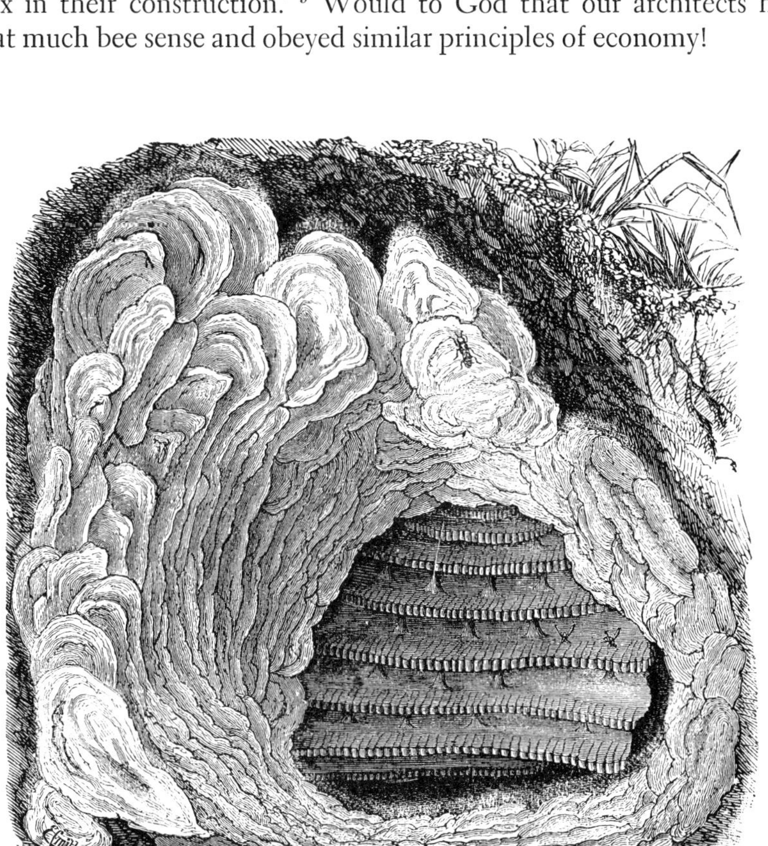

39. *While bees conform to a less-is-more doctrine, wasps are individualists (see also fig. 43). The engraving brings to mind a Diamond Horseshoe seen through a baroque proscenium arch.*

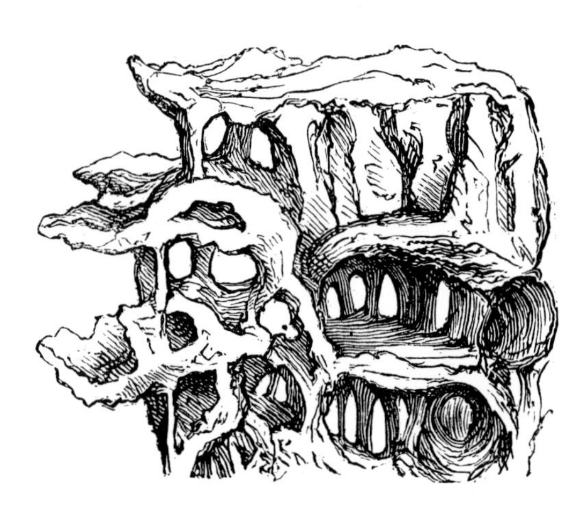

40. A convoluted structure built by red ants.

41. Opposite page: One of several free-wheeling structures which the late architect André Bloc built at Meudon near Paris. (Courtesy Mme. A. Bloc)

In all fairness to man it ought to be said that animals have the advantage of an early start. Some were accomplished builders in time so remote as to defy imagination. Termites, for instance, have been around for more than 300 million years; compared to them man is no more than a straggler in the building field. Yet such is his conceit that he does not blush to compare the termites' constructions to his own. "One would never think that they were not the work of man," writes the Belgian scientist Dr. Deneux apropos the intricate African termitaries; "they resemble banks or mounds of which the inner walls are in the form of spiraling colonnades with a complex system of passages running side by side, passing one below the other, intertwining, and all as regular as if made by machine."[10]

Machine indeed! It is one thing to applaud the termites' performance, it is quite another to compare their work to machine products. Surely, they would decline the left-handed compliment; few man-made buildings come up to their taste, and then only as victuals. Could anybody conceive of a termite so forward as to say of the late André Bloc's hermitages at Meudon, "one would never think that they were not the work of termites." It knows better than that; if anything, M. Bloc's excellent improvisations are comparable to the domicile of *formica rufa,* the red ant.

The termites' houses are remarkable in several respects. Unlike us, termites work without tools. They need no T-square, saw, hammer, or chisel. The upper jaw serves them as trowel, the antennae as rulers. Furthermore, their workers are blind. There may be pianists who can perform Beethoven's thirty-two piano sonatas blindfolded, but one has never heard of a bricklayer or a carpenter working in total darkness. The curious thing about the insects' blindman's buff is the phenomenal durability of the constructions which by far exceed that of our houses. To wit, a large mound near Salisbury in Rhodesia, having been submitted to the most exacting tests, owned up its age of seven hundred years. A pickaxe could not dent it; it had to be blown up with dynamite to make way for a road.[11]

To Henry Smeathman, the father of termitology, the mounds were "foremost on the list of the wonders of creation."[12] This recognition is of long standing; the complexity of the mounds, another admirer, P. E. Howse, reminds us, "impressed biologists of the eighteenth and nineteenth century so much that they assumed without question that the creatures that built them must be highly intelligent and their society comparable in almost every way with human society."[13]

The termites' gothic structures are models of the kind of architecture that our prognosticators are raving about. They loom sky-high, at least in the termites' close-to-the-ground perspective—hermetic containers for a highly compressed inhumanity. One mound alone may

house as many as 3 million termites. But, unlike us, termites are passionately devoted to the common weal. They have been practicing socialism—under a queen, let it be noted—an infinity of time before its doctrine was formulated by man. They distinguish themselves by social behavior, which is more than can be said for the human race. "Within the limits set by the abilities of the species," writes one naturalist, "everything appears to be done for the good of the community and only the necessary minimum for that of the individual."[14]

By temperament and upbringing termites are inveterate overachievers. With their flair for gigantism they easily master structures of extraordinary dimensions. They put up a scaffolding that doubles as a skeleton, and by filling in the empty spaces, they transform it into a solid building. The termitaries of an Australian species, *Nasuthermes triodiae*, reach a height of 26 feet. No wonder termitologists are enraptured with these productions. If termites were the size of men, they argue, their largest mounds would be about four times the height of the Empire State Building and 5 miles in diameter. Which is to say that, proportionally, the termites' undertakings rival—indeed, surpass—the grandest of man's constructions. Although termitic skyscrapers lack stylishness, they sometimes mimic, according to people with a surplus of imagination, "ruined castles or phantasmagoric cathedrals." To more sober observers, their most marked characteristic is the absence of any and all openings. In a way, their mounds parallel today's windowless museums and department stores. We do not know whether the hermitic residences correspond to the insects' actual needs or whether they submit to the dictates of their queen. "Insect societies probably have a language," conjectured Henri Bergson; "by language community of action is made possible."[15]

Actually, it is less the bigness of the termites' high-rises than their flawless technical organization that commands our respect. To insure the health and happiness of a mound's teeming millions, they keep the mounds at an even temperature the year round by installing hundreds of ventilating ducts that reach into the farthest corners of a structure while fermenting grass and debris generate smokeless heat. The cells, halls, and corridors are kept spotlessly clean at all times and are continually repaired or rebuilt. Waste disposal, including funerals, presents no problem. Instead of burying their dead, the termites eat them—an expedient far in advance of other societies. All surplus population becomes food. Maeterlinck, who wrote an exceedingly detailed account of the termites' life, thought their civilization "not a whit inferior to the civilization we are attaining today."[16] It is, he maintained, the earliest of any, the most curious, the most complex, the most intelligent, and, in a sense, the most logical and best fitted to the difficulties of existence that has ever appeared before our own on this globe.

Termites are ahead of us in other respects. To avoid traffic jams they invented one-way streets. They merged the production of shelter and food by building edible houses, a type encountered on the human level by Hänsel and Gretel only. In years of plenty, trees and wooden buildings are their daily bread. When threatened by a wood shortage, they consume a mound's walls, which are largely made of fecal matter—a splendid example of recycling. No doubt, the termites' robust health and longevity are due to their cellulose diet, balanced occasionally with leaves and fungi. A mass of protozoa in their intestines make, like so many cooks, the raw material digestible for them. Is it too farfetched to assume that some day also our stomachs could be converted to cope with a ligneous diet? That instead of bulldozing a derelict bungalow into dust, it will be cut up into bite-size morsels? With the sources of our traditional foodstuffs becoming rapidly depleted, we may very well conceive of a family, not giving thanks over a roast turkey, but feasting on a well-seasoned length of two-by-four. By coaxing our enzymes to slightly rearrange the chemistry of our alimentary tract, we might dine as fastidiously as the lowliest termite. Maeterlinck probably had prospects like these in mind when he called termites "the heralds, perhaps the precursors, of our destiny."[17]

If termites presage a homogenized humanity, birds represent the incorrigible mavericks. Birds' nests, which are masterpieces of compressed composition, probably did not go through many evolutionary stages but emerged early in definite form.

42. Swallows build their nests with moist earth and straw, the same materials that Orientals use to make wall surfaces of singularly attractive texture.

To those of us fortunate enough to have had an entrée to Nature in our youth and to cherish recollections of going nesting, the word *nest* conjures up a frail little cup, woven of grass, bits of leaves, feathers and fluff, *bosomed* into shape. "A bird's tool," wrote the historian Jules Michelet, "is its own body, that is, its breast, with which it presses and tightens its materials until they have become absolutely pliant, well-blended and adapted to the general plan."[18] The classical receptacle for eggs and baby birds, the nest symbolizes love and tenderness, "*un bouquet de feuilles qui chante*."[19] (Naturalists do without such poetic notions. They speak of reptiles' and fishes' nests, and refer with equanimity to a 450-pound gorilla's lair as a nest. Although the simian product does not resemble any permanent human habitation, it is the equivalent of our house-within-the-house, the bed.)

In zoos animals have proved their ability to survive under most trying conditions. Solitary confinement and the torture of caging may shorten their life span and dull their senses but do not necessarily cause illness. What we do not quite know are the difficulties that animals which escaped the cages of zoos and circuses had in readjusting to their original habitat. Or, for that matter, what noticeable effect on his daily way of life have man's periodical escapes to his vacational camping grounds. "In many respects," observed René Dubos, "modern man is like a wild animal spending its life in a zoo; like the animals, he is fed abundantly and protected from inclemencies but deprived of the natural stimuli essential for many functions of his body and his mind."[20]

Like humanity, the animal kingdom is by no means short of irrational types; it, too, has its quota of maladjusted individuals. Take the hammerhead, a singularly insecure African bird. He builds a veritable bunker, oblivious or contemptuous of avian etiquette according to which a bird's nest ought to be inconspicuous. Although the hammerhead is no bigger than a raven, his nest measures 6 feet across. Paradoxically, it has but a tiny hole for an entrance. The perfect psychotic touch is provided by its flat-topped roof, strong enough to support a man's weight. What antediluvian fright makes this bird seek safety in a fortress?

The other extreme in the way of an avian lair is represented by the stork's nest. Open to the elements, precariously, not to say, foppishly, perched atop a chimney or a spire, it is little more than a landing pad. The loftiness of such premises makes one believe that a stork is a supremely balanced bird. Which, of course, he is, if only physically. For he can sleep standing up on one leg, thanks to a locking device in his knee and a rare sense of equilibrium. A pathological variant of the stork's nest is the stylite's perch. Although it does not enclose space, it has to be classed as habitation, if ever so ascetic. Its inventor, the Syrian monk S. Simeon Stylites, lived on a platform 3 feet wide and 18 feet above ground. He later raised it to 60 feet and stayed there in relative comfort for thirty-seven years.

In view of the great variety of birds' nests it ought to surprise no one that there are birds that like to live in the avian equivalent of our high-density housing projects. So do, for instance, the social weaver finches of South Africa. These birds immoderately enjoy each other's company and carry out their building projects in a communal spirit. After collecting wagon loads of grass, they make a roof in a tree under which each pair puts up its own cubicle, cozily lined with fluff. And such is their gregariousness that a single one of these structures shelters as many as three hundred couples.

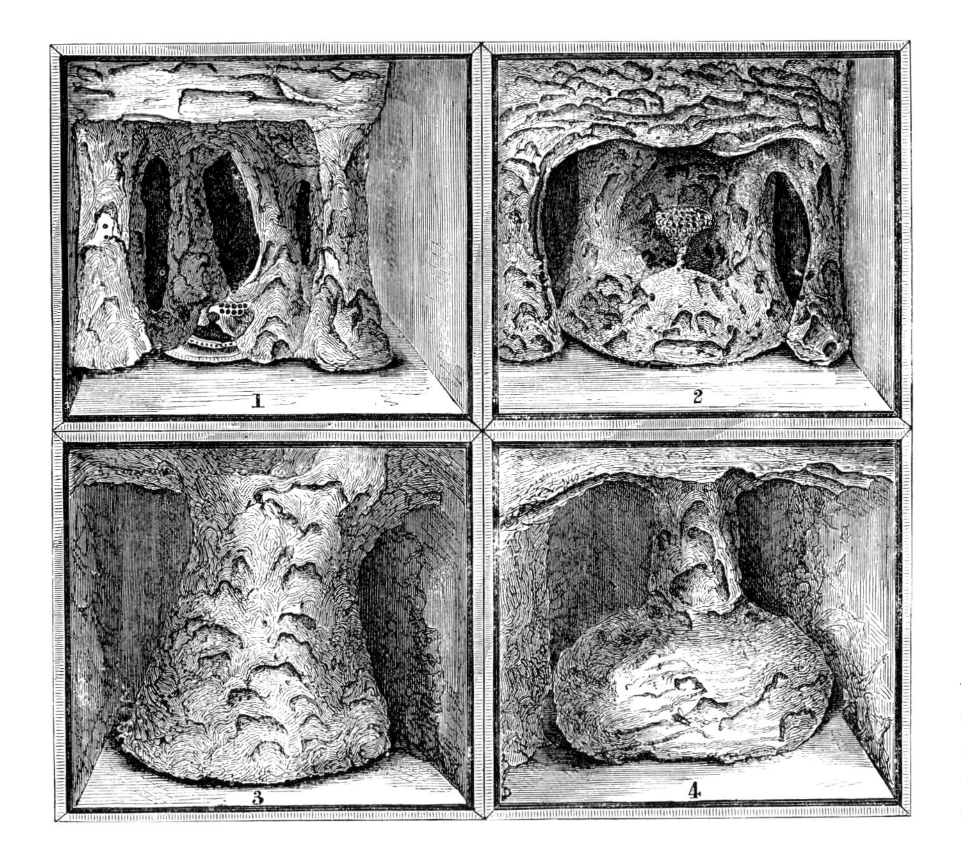

43. An English naturalist, desirous to test the inventiveness of wasps, induced them to build nests in square boxes. The results were thought remarkable enough to be enshrined in the British Museum.

We speak of love nests and human warrens, and although words like *foxhole* and *crow's nest* have entered military and nautical terminology, our domestic architecture owes little to the animal kingdom. (Philosophers ridicule our mania of injecting human qualities into nest building; "among birds," Bachelard pointed out, "love is a

44. Weaverbird and his nest in the shape of a retort. From Voyages aux Indes Orientales *(1776) by Pierre Sonnerat.*

strictly extracurricular affair, and the nest is not built until later, when the mad love chase across the fields is over."[21])

If to the termites can be ascribed the invention of towering structures, to the weaver birds must be given credit of having invented basket work. It isn't so much the shape of their nests that distinguishes them from their fellow birds, even though some of them are fairly extravagant; what puts them in the forefront of expert artisans is the downright human touch of their performance. Or should we, who have been lagging 30 million years behind our feathered friends, not rather speak of our avian skill when putting up wattle-and-daub? The fact that a number of beasts are competent in all sorts of building trades, and have been so long before man was able to stand upright, ought to blunt the edges of our presumptuousness.

We have no means to determine how weaver birds arrived at their curious nest-making technique. Their nests are not conventionally shaped like cups but are made in the form of a retort or kidney, to be entered from below or at the side. The all-round enclosure not only protects them from wind and rain but also from mosquitoes whose bite causes them malaria. It is the male's business to build the nest while the female provides the floor covering for the egg chamber. "The most remarkable feature is the outer basketwork, a true fabric of great strength and pliancy," notes ornithologist John Hurrell Crook; "the woven fabric is the result of a finely developed stitching technique in which strands of material are inserted, recovered and reinserted in several ways."[22]

Building materials include strips of grass or palm leaves—weaver birds are Africans—tendrils, needles, or any suitable fibers. These are knotted, entwined and interlocked as artfully as if worked by ten fingers. Yet the bird's lack of appropriate extremities for what can be called a handicraft without hands does not impair its performance. "Either one foot or both hold the grass strand," explains Mr. Crook, "the end of which . . . is pushed around to the far side of the twig, whereupon the grass is seized again, inserted below the strand, pulled tight, and once more pushed out on the far side of the twig."[23] All this is achieved with its beak, without knitting needles and crochet hooks, without the ghost of a loom.

The bird's work speed, too, is a cause for wonder. The construction of a nest, begun at 8:30 A.M. (clocked in by Crook), despite the visit of a female and an ensuing courting episode, was finished, or almost finished, by 5:30 P.M. The nest only lacked the inner lining, which was added the following morning.[24]

And what is one to make of the bowerbirds' capers which so delight and mystify the ornithologists! Defying all our notions of birds and their brain's capacity, the bowerbirds exceed their nest-

P. Sonnerat Pinx — *J. J. Avril Sculp*

building instinct by arrogating themselves spheres of activity which in human society are reserved for decorators and stage designers. These birds, who are natives of New Guinea and parts of Australia, and whose showoffishness surpasses that of their relatives, the birds of paradise, enthusiastically practice conspicuous display. In order to conduct their courtship according to bowerbird etiquette, they construct a playground, or dance floor, which they adorn with moss and flowers—preferably orchids (to be replaced by fresh ones as soon as they wilt)—with feathers and colorful berries, but also, true to the decorator's calling, with *objets trouvés* such as shells, bones, and bits of colored glass. Endowed as they are with a pronounced color sense—the blue males collect blue flowers and blue glass while the green females are partial to greenery—they don't stop at flower arrangement but take off into the higher realm of art. "These highly decorated halls of assembly," wrote a bowerbird watcher, J. Gould, "must be regarded as the most wonderful instances of bird architecture yet discovered."[25] Ages before artists learned to grind their pigments, bowerbirds used vegetable dyes to paint their bowers. A beak, dripping with fruit juice or fruit pulp, is as good as a paint brush.

The invention of the bird cage must have been sad news for the feathered world. The cage is not a house but a prison which comes close to reflecting our own paltry concepts of comfort. A crib for feed, a cup for water, a bath, and a swing roughly correspond to our own lowest level of subsistence. Now it is true that domesticated animals are not choosy and accept unprotestingly what their master provides. No cow ever had its own ideas about what makes a stable, still less a desire to build one. The same holds true for dogs. They have no complaints about their doghouses, being unaware of the word's derogatory implications. Yet, passing in review all the kennels, coops, hutches, and sties, one cannot but feel pity for their inhabitants. None of their carpentered lodgings come up to those sophisticated ones built by self-sufficient beasts. Hence when looking at animal architecture we can best understand urban man's irresolution and ceaseless experimenting when faced with building a shelter.

45. A structure put together with the bones of animals. From Klemm's Allgemeine Culturgeschichte der Menschheit.

46. To judge by the artistic level of weaving, painting, and decorating birds, some species may be as keen as man on selecting a becoming background. These seabirds in the Faeroe Islands have about the same taste for theatrical scenery that made monks and monarchs choose near-inaccessible sites for their preferred abode.

The only presentable man-made animal house is the dovecote. Far from confining the birds, it encourages their independence; every day is open house. From the point of view of looks, the dovecote is in a class of its own and merits more than a cursory glance. Like so many other exemplary utilities, it does best in the Orient. For what *we* call a dovecote is but a paltry version of the real thing. However, before entering into the particulars of its physical shape, a biographical note on its tenant is in order.

To drive home the dove's importance in southern and eastern countries—in northern ones it has practically no standing (aside from its symbolic significance)—it will be necessary to briefly examine its singular position in profane and sacred history—from the vermin-infested bird begging for crumbs on its favorite beat, the cathedral square, to that very same cathedral's august tenant, the zoomorphic Sanctus Spiritus, the Holy Ghost. It always pays to learn the basic facts about the *inhabitant*, whether human, bestial, or divine. Historiographers rarely fail to acquaint their readers with the lives of patrons of the arts, down to the most intimate details; it is only just to give equal time and space to pigeonry, whose most exalted member, the God-pigeon, ranks above popes and potentates.

47, 48. Doves do not seem to be committed to any particular dovecote style, as the two strikingly dissimilar versions from Greece and Spain suggest. The one from Mykonos (opposite page) radiates the charm of authentic folk art, while the one from Barcelona (right) has its roots in the urban vernacular.

"What a wide gulph separates the Pigeons from all other captive and domestic birds! How completely discrepant are all their modes of increase and action, their whole system of life, their very mind and affections,"[26] wrote one Reverend Dixon in his book *The Dovecote and the Aviary* more than a century ago; "they have created as types almost Christian virtue."[27] Yet Dixon was too much a man of the world not to allow for occasional dovish capers. Speaking of widowed doves, he asserts that "*femmes seuls* intending to keep so, are things intolerable in a columbine society."[28] The realistic *Reallexikon*, too, pooh-poohed the popular belief which attributed purity, chastity, and marital fidelity to the dove before Konrad Lorenz let it be known that pigeon and dove can be savagely aggressive. Nevertheless, the bird is loaded with noble symbolism. From Noah to Picasso it has stood for hope and peace. (In common usage *dove* is the poetic designation, *pigeon* the prosaic one—a fair distinction for a bird that is at once divine and exceedingly mundane.) Long before being canonized by the Church, it was worshipped throughout the Old World. In Assyria and Babylonia it was the attribute of Istar, goddess of love and fertility. Perversely, it also embellished the armies' banners. At Hierapolis, the center of the dove cult, a golden dove crowned the head of the great nature-goddess Atargatis. Soothsaying wild doves nested in the branches of Zeus' old oak tree at Dodona, doves and dovecotes abounded in the temple of Venus in the town of Paphos on Cyprus. In Olympian circles the dove had the status of a pet and traditionally sat on the shoulder of the Naked Goddess (she did, however, wear shoes), serving her as a godly attribute and, presumably as a surrogate of dress. One encounters doves in hallowed Punic texts, and as sacred birds of the Ka'ba, Mecca's holy of holies. Syrians and Assyrians refrained from eating doves because they held them sacred, and so do today's Muslims. Doves rate low with us; not by accident did the Americans appoint a bird of prey, the eagle, as their national symbol.

Christianity took no offense at the bird. On the contrary; as with so many heathen assets, it incorporated it into its realm, allotting it the highest honor, a share of the Trinity. (By definition, the Trinity is "distinct from, but consubstantial, co-equal and co-eternal with, the Father and the Son, and in the fullest sense, God."[29] Tertullian spoke of the Church as *columbae domus*, the Dove's House.

The houses of the profane dove are a different matter altogether, particularly in the Near Orient where they sometimes eclipse in size and number all human dwellings. "You find here more dovecotes than other houses,"[30] wrote Maundrell, an eighteenth-century traveler to the Holy Land, and even in our days one still comes across veritable pigeon towns. In Persia, Turkey, and Egypt, pigeonry's homelands par excellence, dovecote architecture is in a category all by itself. Although the birds' demands are similar in all three countries, their accomodations differ widely. In Persia they may dominate an entire landscape. Of the more than three thousand dovecotes once concentrated around Isfahan, a good number are still in working order: stout towers, 40 feet high, which the uninitiated may easily take for fortifications. They consist of concentric cylinders, the inner one built higher than the outer one and topped by an extra turret. Since building or owning a pigeon tower was denied to Christians, those who were eager to obtain the privilege turned apostate and became Muslims, to the chagrin of the Holy Spirit.

It isn't that all these people subsist on a diet of squab; what they covet are the birds' droppings. The excrements accumulating inside the towers make precious manure, and pigeons are solely bred toward that end. The cotes' importance for the country's economy is driven home by the fact that kings once levied a tax on the dung. (By the same token, emperor Vespasianus imposed a duty on the collection of

49. This seventeenth-century print shows three of the three thousand–odd pigeon towers that once dotted the countryside around Isfahan. Although they were uniformly plain, their battlements left some margin for decoration. From Sir John Chardin's Travels in Persia *(1668).*

the urine produced by Rome's citizenry, a shrewd stroke of business sense that earned him lasting monuments in the *vespasiennes*, the public urinals that grace Paris's streets.) Solid as the Persian towers appear to the eye, they are nevertheless vulnerable, not so much to high winds but to gusts generated within. A snake's sudden appearance in a tower will spread panic among the doves, and the vibrations set off by the whirrings of their wings may cause the wall to split.[31]

Egypt's agriculture is similarly tied to a dove culture which goes back to about 3000 B.C. "Pigeon-fancying is carried to insane length by some people," wrote Pliny in the chapter on Egypt in his *Natural History*; "they build towers on their roofs for these birds and tell

50. Pigeon towers of jagged silhouette rather than human dwellings add a note of distinction to the homely villages along the Nile. From Le Brun's Voyage au Levant *(1700). (Courtesy British Museum)*

stories of the high breeding and pedigrees of particular birds."[32] More than fifteen hundred years after Pliny, Paul Lucas (whom we met in a preceding chapter) found the pigeon economy unchanged. Almost all houses, he wrote, are topped by pigeoncotes in the form of square towers with crenellations, painted red and white.

As in Persia, in Egypt pigeoncotes nobly stand out from the domestic architecture. "Traveling up the Nile Valley by rail," one traveler wrote, "the many pigeon houses to be seen in the villages and on the outskirts of the towns passed en route are generally the most arresting buildings noticed on the journey."[33] Although they are not counted among the wonders of the world, they are visited by travelers

à titre de curiosité. The dovecote of prince Youssouf Kamal at El-Qasr sheltered several hundred thousand wild birds simultaneously. When they took flight, it seemed that night had fallen.

Whereas our urban pigeons are about as useful as goldfish, to the farmer who lives in the semideserts of the Near East doves are his life insurance. Since cattle dung is used for fuel, orchards and gardens have to be fertilized with pigeon dung. To assure successful crops every year, farmers depend on a never-ending supply. The birds, Sir James Richards noted, are the symbol of agricultural stability; "it is impossible for a young man to look for a bride unless he possessed at least one pigeoncote."[34]

Pigeoncotes reach their apotheosis in the Turkish province of Cappadocia. For generations travelers have stood before this architecture in the raw, lost in profound astonishment and unable to satisfy their

51. In Egypt, pigeon dung has been, and still is, the mainstay of the country's agriculture. In return for their services, the birds are allocated veritable mansions which are, architecturally speaking, several cuts above the looks of human dwellings.

curiosity. And it is true today as it was in the past that even any halfway explanations are hard to come by. Historical accounts are scarce, the peasants uninstructive, and government officials noncommittal. Although the Cappadocian pigeon territories are now being groomed into showplaces for the tourist industry—motels and asphalt roads are all that is missing to put them on the travel agent's map—the relative inaccessibility of the pigeoncotes helps to perpetuate the mystery.

"We saw pigeons flying out of the upper openings to which there appeared to be no external means of approach," wrote one William J. Hamilton, secretary to the Geological Society, who visited the region in the 1840s.[35] What intrigued him then, and what equally intrigues today's traveler, is the fact that the rock surface around these openings is carefully smoothed, decorated with colorful ornaments and

52. *Anatolian pigeoncotes, carved into a perpendicular mountainside, achieve some of the nobility of royal rock tombs.*

sometimes inscribed with Greek letters. A climbing tour—or rather a roping down—for a close view is not advisable. The porous stone wastes under the beatings of wind and weather, and rockslides are common. From time to time 100-feet-high façades are dismantled, as it were, by invisible demolition squads, revealing shallow interior rooms whose walls are honeycombed by row after row of niches. Obviously, the niches are pigeonholes—but have they always been? Moreover, who were the men who braved the perpendicular cliffs to carve the thousands of niches and mark them with lettering and paintings? Are we not in the presence of abandoned burial grounds that subsequently were converted into breeding grounds? Or did the pigeons precede the dead?

The confusion is compounded rather than cleared up by etymology. *Columbaria* denoted both pigeoncotes and poor people's graves. In antiquity, graves having been as costly as they are today, undertakers who speculated in underground real estate found a way to save precious space by arranging cinerary urns in rows, like jars on a larder's shelves. This kind of dead storage is still customary in southern countries. The question of which came first, pigeonholes for the dead or pigeonholes for the birds, seems as moot as the one of the precedence of chicken and egg.

53. *Opposite page: This apocalyptic scenery resulted from rockslides that bared a warren of columbaria, depositories for human bones and pigeon droppings. Göreme Valley, Cappadocia, Turkey.*

54. *A tiny cross atop the four-tiered columbaria at left is the only clue to the nature of these buildings—silos for the dead.*

The other question—who were the original inhabitants of the rupestrian dwellings: doves? men? or both?—finds an answer of sorts in the Scriptures. In biblical times urban life seems to have left as much to be desired as in ours, and to escape to the country was one of the remedies. The panegyrics of our real estate developers had their counterpart in the entreaties of the prophets. "O ye that dwell in Moab," touted Jeremiah, "leave the cities, and dwell in the rock like the Dove that maketh her nest in the sides of the hole's mouth" (48:28).

These conjectures by no means exhaust the theories about the dovecotes' original function. There are people who believe that they once were inhabited by the same troglodytes we encountered on page 33, men and women who enjoyed living with the free-wheeling birds as subtenants. But the innate shyness of doves makes this seem improbable. On the other hand, it is true that in better times the sharing of common quarters by man and beast was nothing unusual.

55. The stark scenery, the habitat of wild doves and deceased monks, makes our aviaries and cemeteries look banal by comparison. Near Maçan, Anatolia.

The boldest thesis attributes to the wall dwellings a triple function: abode, cote, *and* tomb. Strange to say, feeding troughs of stone were found in abandoned cubicles high above ground. "Here," the above-mentioned Hamilton noted, "the mangers are hollowed out of the rock, and may almost be mistaken for ancient tombs or receptacles for sarcophagi or urns."[36] If we have here to do with mangers, so far nobody volunteered a theory of how the animals managed to get up to the dizzying heights. It seems more likely that the mangers may have been sarcophagi. Since intimations of death are ever-present in ancient domestic architecture, the thought of sharing a room or house with the dead may not have struck the inhabitants as macabre.

Moreover, dovecotes and tombs have been intimately connected for ages; on tombs and sarcophagi the dove is often depicted as symbol of the departed's hope for peace and resurrection. Let us take leave of the pigeons and their houses on this optimistic note.

56. Before the world was scrutinized by the cold eye of the camera, artists rarely missed a chance to improve upon reality when doing an architectural rendering. It is doubtful, though, that the builders of Stonehenge would have been flattered by this eighteenth-century representation of their handiwork. (*Courtesy British Architectural Library*)

When architecture was all play and no work

"Archaeology," maintained Sir Flinders Petrie (1853–1942), the eminent English Egyptologist, "gives a more truly liberal education than any other subject as at present taught."[1] His compatriots may be unaware of his dictum but, luckily, they have something like an innate affection for antiquities, coupled with a magpie complex. Although serious collecting is the pastime and privilege of the rich, the little man occasionally likes to give himself airs and play the connoisseur. On holidays he is not averse to strolling through the collections of the local museum or to joining a lecture tour among some musty ruins. Besides, every year hundreds of thousands of people travel to Stonehenge to have a look at what is considered the most important antiquity in the British Isles. To call on this venerable pile of stones has become as much of a ritual as visiting Niagara Falls and the Grand Canyon. Indeed, Stonehenge impresses many visitors as a physical prodigy rather than a cultural monument, perhaps because the monoliths have not been overly dressed. If anything, the uncouth stones in a featureless landscape suggest a geological outcrop. Unstandardized as they are, they show no resemblance to classical columns which are forever uniform, like chair legs turned out on a lathe. However, for all its majesty and rude stylishness, Stonehenge does not immediately announce man's triumph over matter.

The present-day fascination with this dinosaur of architecture has not always been shared by past generations; they could never quite decide whether the megalithic builders constituted an old guard or an avant-garde. For whatever may be the proper qualifications for a national monument, Stonehenge falls short of them. "A product peculiarly British,"[2] it badly lacks pedigree. Its credits are minimal. Try as one may, no dates or names can be attached to the stones; they have remained stubbornly anonymous. Neither Tacitus nor Caesar mention them in their descriptions of religious observances in Britain, and although the Romans paid homage to lesser world wonders, they seem to have overlooked Stonehenge. So did, by the way, the Saxon chroniclers. To live down these snubs, patriotic Englishmen did not blush to disallow any great age to the stones. They preferred to believe that they had been put up *after* the departure of the Romans, which is all the more ironical as Stonehenge antedates by centuries the cyclopean walls of Mycenae, once believed to be Europe's farthest architectural outpost in time.

Attempts to enroll Stonehenge among Architecture's nobility were made from time to time with varying success. Yet, unlike classical monuments, Stonehenge remains a subject of speculation to this day. Is it a clock? a theater in the round? a hustings? an architectural tournament? an incantation? Educated guesses and semiliterate musings, patriotic fabrications and old wives' tales, digs, surveys and, lately, radioactive-carbon dating have only compounded the mystery. For a pile of old stones that do not even carry inscriptions, the notices it has accumulated are impressive. To mention but two amateur archaeologists who probed Stonehenge in early times: Bishop Geoffrey of Monmouth, the twelfth-century historian and fantasizer, identified it as a war memorial to those fallen in the battle with the Saxon leader Hengest. It is a fairly cautious guess and not quite as unreasonable as it might seem because war monuments do not follow any established style. Far bolder, and probably more ingratiating, was a theory advanced in 1624 by the poet Edmund Bolton. He added centuries to the age of the stones, and more than a touch of national prestige, by putting them down as the burial place of the British Queen Boudica, alias Bodicea, a contemporary of Nero's. To clear up a matter which increasingly occupied the minds of scholars and laymen alike, King James I commissioned Inigo Jones, the British Palladio, to look into the stones' origin.

As it happened, Jones's opinion of native architectural relics was low. "Touching the Manner of the Buildings of the ancient Britains and of what materials they consisted," he wrote, "I find them so far short of the Magnificence of this Antiquity, that they were not stately, nor sumptuous; neither had they any thing of Order or Symmetry, much less, of Gracefulness and Decorum."[3] These shortcomings he set out to correct.

Two journeys to Italy had left him overly temple-conscious and, intent as he was on ennobling England's anonymous relics, he declared Stonehenge to be nothing less than the finest fruit of architecture, a temple. Although he had thought the stones to be Druidic, he dismissed this first intuition and settled for a more prestigious origin by resolutely diagnosing them as the ruins of a Roman-built edifice. Letting himself be carried away by his zeal for certainty, he further succeeded in discovering that the temple has been dedicated to "Coelus, the senior of the heathen Gods, and built after the Tuscan order." To reinforce the nation's faith in his disclosures, Jones wrote a learned dissertation on the subject and published it, complete with a reconstruction of the monument. The illustration on this page is taken from its pages. More than a century after Jones, the antiquarian William Stukeley, an early field archaeologist of a highly romantic bent, revived the Druidic hypothesis in his celebrated work *Stonehenge: A Temple Restored.* Due to his and similarly indelible writings, a neo-Druidic vogue was promulgated and, to this day, bogus Druidic rites are annually performed on the Stonehenge grounds.

Other men's interpretations—triumphs of inconsistency—vacillated between cenotaphs and various types of sacred enclosures, tombs, a fortress, a theodolite (which is an instrument for measuring horizontal and vertical angles), and a sanctuary to the moon. None of these theories was regarded as entirely satisfactory. In 1927, one A. P.

57. *This fanciful reconstruction of Stonehenge is taken from Inigo Jones's treatise on the controversial monument, written at the command of James I and published in 1655.* (*Courtesy British Architectural Library*)

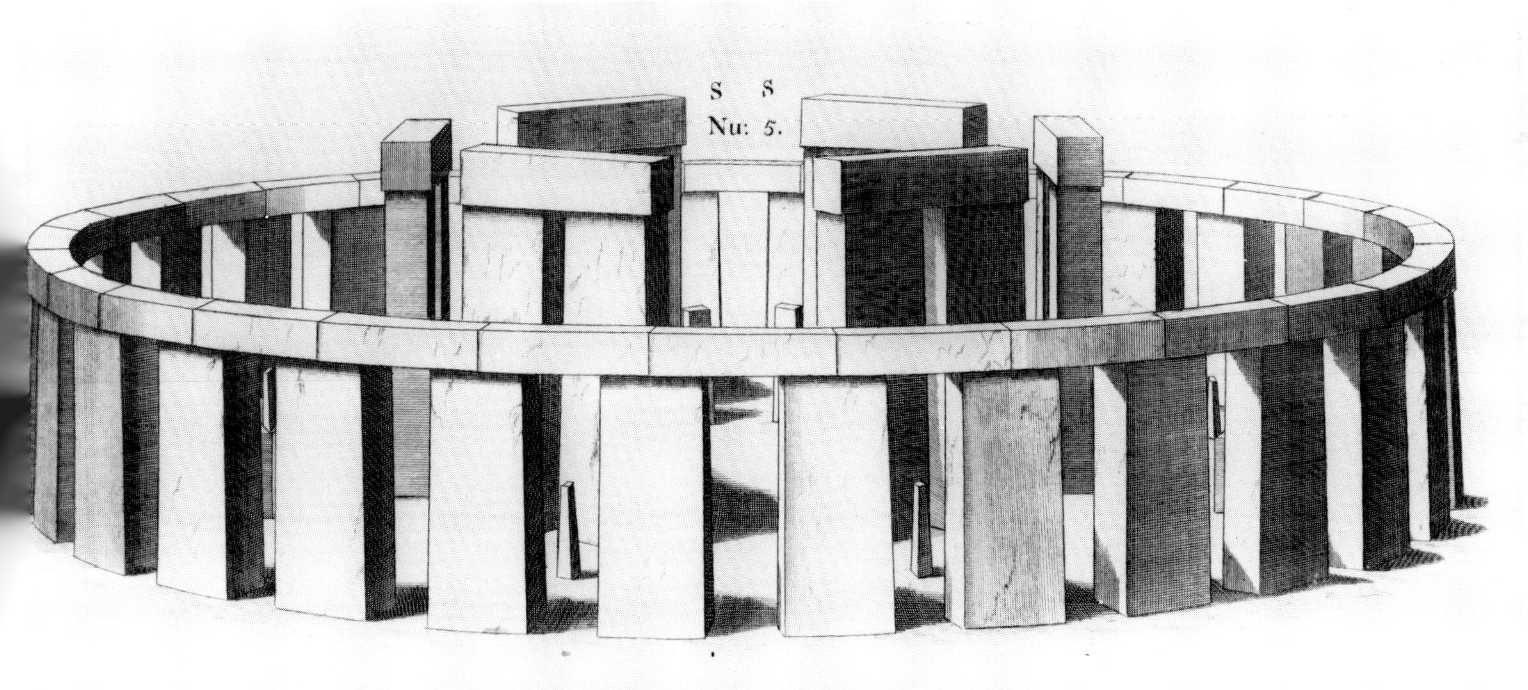

Trotter came up with a suggestion that was to be taken up half a century later: Stonehenge was an astronomical instrument. It consists, Mr. Trotter declared, of a pair of "sights" like those of a rifle. "The fore-sight is the pair of stones forming the entrance. The back-sight *was* the pair of stones forming the largest trilithon."[4]

The builders of this multipurpose edifice were believed to have been of all hues, not excluding the distant Phoenicians. Since no risks were involved in the guessing game, it was played with more gusto than discernment. The more meager the architectural remnants, the more they are apt to stimulate the imagination of the would-be restorer. A latter-day theoretician, one Christian Maclagan, perhaps with Rome's Castel Sant'Angelo in mind, convinced himself that the stones had been "enclosed in masonry with a domed roof, the whole structure forming a military defense work."[5] Stonehenge bravely rode out the sundry theories that were hung on its skeleton.

Above the muddles of these ever so nobly motivated Englishmen sounds the clarion voice of one W. S. Blacket, American patriot and dauntless amateur historian, who had arrived at stunningly original conclusions without agonizing and hairsplitting, guided solely by the texts of the ancient authors and the sort of straightforwardness that distinguishes the sons of his nation. In his modestly titled work *Researches into the Lost Histories of America*, published in 1883, he solved the thousand-year-old riddle to his and his compatriots' satisfaction: Stonehenge, he asserted, had been built by American Indians.

The acute observer that he was, and open-minded enough not to take amiss subversive religious practices of the past, he decided that Stonehenge had once been the house of a pagan god. This added nothing new to current opinions; where he departed from his predecessors was in the exactness of his revelations. "The great and wonderful erection," Blacket announced with evident relish, "which stands on the plain of Salisbury—a puzzle to antiquarians and a mystery in the obscure history of Great Britain—was a temple of Apollo."[6] The shocking part of his discovery was that he dared to dispute the British origin of the monument. "It is quite manifest," he maintained, "that Britain could not have received the religion of Apollo from Greece"; it was an American import. Britain, he assured his readers, got it via the New World, for "the religion of the European Apollo must have had its rise in the southern mound cities of the United States."[7] (Englishmen, knowing next to nothing about American mound cities, may have found this news hard to disprove.) Warming to the subject, Blacket concluded that "it must therefor be inferred that the Appalachian Indians with their priests and medicine men must have been the builders of Stonehenge."[8]

For Blacket everything tumbled into place. "To find the birth of

Apollo in the mound cities of Northern America," he explained, in his tousled prose, "is to supply a natural and credible explanation of the mystery of Stonehenge. The Appalachians are said to have erected altars and to have massed heaps of stone wherever they founded colonies, and they may have carried with them the religion of the race."[9] British pride may have balked at the thought that England had once been an American colony, but historians will find it irksome to explain away the savagery with which British soldiers fell upon the Appalachians in the early eighteenth century. Having defeated them in battle, they killed their priests, annihilated the tribe, and sold what was left of it into slavery.[10]

At this point an extended footnote on the mound cities whence, according to Blacket, hail the builders of Stonehenge, is in order.

The designation is misleading. The tourist bent on exploring the ruins of North American prehistoric cities is in for a disappointment. Nothing as grandiose as Mexico's archaeological sites has turned up or is likely to emerge in time. The mounds in question are earthworks, a sort of architecture only slightly less ephemeral than sand castles. Their builders, the Indians, kept them in good shape, but after their untimely departure they crumbled away. The first white settlers who filled the vacuum may have come upon some mounds in all their glory, only to plough them up, level them, or to cap them with their unlovely houses. Nevertheless, a handful of men did recognize their significance and set to recording the sites with commendable precision. In most cases, descriptions and survey maps are all that is left of the ancient monuments.

At the turn of the century no more than one-fifth of the mounds and earthworks that once had covered the broad valley of the Mississippi River could be identified. The growth of St. Louis had destroyed the last vestiges of a large group of pyramids, terraces, tumuli, and "falling gardens." Still, what remained of them was impressive. "When I reached the foot of the principal mound," related one Brackenridge who visited the antiquities of the Mississippi basin in 1811, "I was struck with a degree of astonishment, not unlike that which is experienced in contemplating the Egyptian pyramids. What a stupendous pile of earth! To heap such a mass must have required years, and the labor of thousands. Were it not for the regularity and the design manifest, the circumstances of its being on alluvial ground, and the other mounds scattered around it, we could scarcely believe it to be the work of human hands."[11] The reiteration of disbelief drones like a basso ostinato.

The truncated pyramid of Cahokia that Brackenridge beheld was indeed of imposing dimensions. It rose over a parallelogram of 700 by

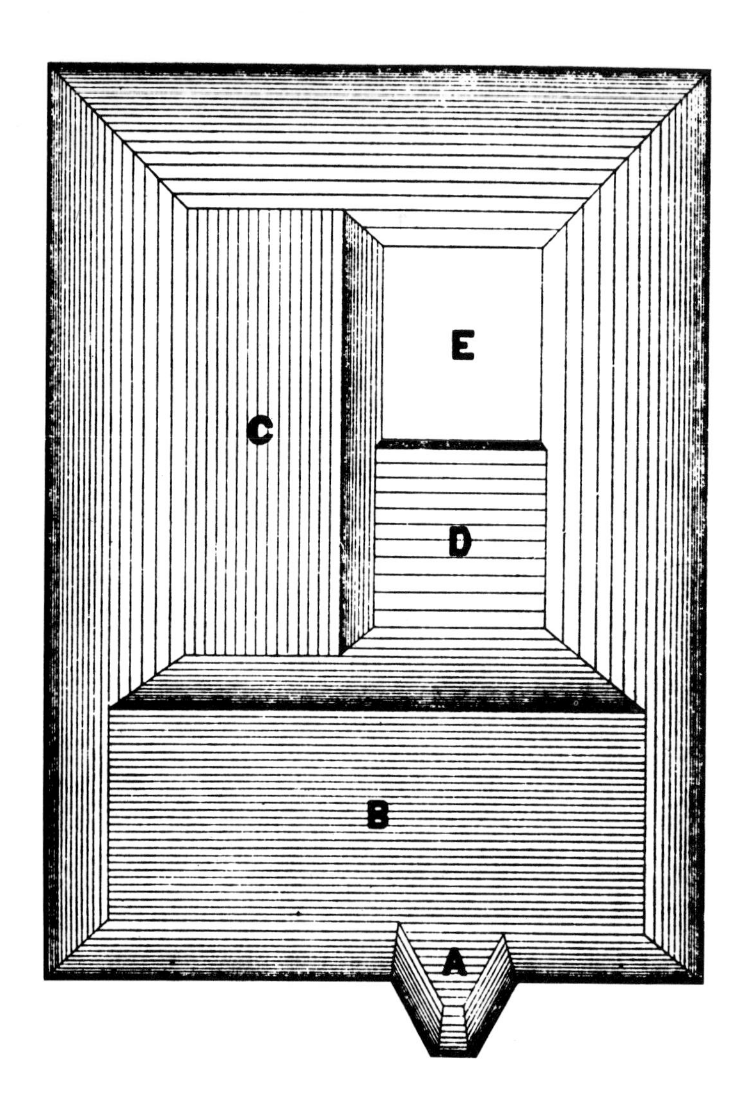

58. Prehistoric Cahokia Mound near St. Louis, the largest of truncated pyramids in the United States, once rose in five steps to a height of 90 feet. As it is entirely made of earth, the sharp contours indicated in the diagram have long been eroded.

500 feet to a height of 90 feet, the summit taking up several acres. The fact that its terraces had been turned into gardens did not lessen its stateliness. At any rate, Brackenridge needed no searching or digging to support his discernment. Being prone to envisioning urban centers where the eye perceived but a boundless prairie (a phenomenon comparable to that of the desert traveler who sees oases where there are none), he concluded that "a very populous city had once existed here, similar to those of Mexico, described by the first conqueror. The mounds were sites of temples, or monuments to the great men."[12]

The "great men" were equally subject to wild surmises. The less known about them, the larger they loomed. As so often, patriotism, with a dash of racial prejudice, played its part in obfuscating the issue. The boldest of the theories attributed the Mississippi vestiges to "a colony of Welsh, or Danes, who are supposed to have found their way by some accident to this country, about the ninth century."[13] Men parched with thirst for a heroic past will not shrink from fabricating instant myths.

Brackenridge's arguments probably seemed plausible enough to his fellow citizens, the more so since they got the backing of like-minded would-be savants. "After many days of exploration and study," one William McAdams wrote apropos some other American antiquities, "we believe the evidence to prove this to be a group of the greatest mounds on this continent and perhaps in the world, and possibly this was the Mecca or great central shrine of the mound-builders' empire. Upon the flat summit of the pyramid, one hundred feet above the plain, were their sanctuaries, glittering with barbaric splendor, and where could be seen from afar the smoke and flames of the eternal fire, their emblem of the sun."[14]

Such enthusiasm was contagious. At long last, North Americans had found their own Pyramids. But who had been the pharaohs? The unglamorous Danes certainly did not fill the bill. If the discoveries of antiquities were to rival those of the Old World, a more venerable race of builders had to be ferreted out.

The discoverers who splashed their rapturous prose over monuments invisible to the naked eye were equally adept at genealogizing.

59. This scene, reminiscent of a music hall spectacle, is set on an unidentified circular mound that might have been created by a pastry cook rather than by prehistoric Indians. Both mound and battle are bogus, dreamed up by a zealous would-be historian. From Traditions of the De-coo-dah *by William Pidgeon.*

Preoccupied as Americans were then with more pressing problems, they wasted no time in taking stock of the continent's penumbral past. Hence Phoenician inscriptions were being deciphered forthwith on Massachusetts rocks, Egyptian mummies unearthed in a Kentucky cave.[15] One William Pidgeon offered "conclusive evidence of Roman and Grecian populations" not only within the country's borders but far to the south. "North and South America," he wrote in a weighty book, "were not only known to the Romans and Grecians, but were formerly taken possession of, and colonized by them."[16] Another run for a respectable ancestry was on.

This pathetic urge to improve upon the pedigree of both natives and newcomers goes back centuries. Men with a sprinkling of classical education, ill at ease among the noble savages and the rabble of settlers, found relief in wishful thinking. "I am bold to conclude," wrote one Morton in his *New English Canaan* (1637), "that the originall of the Natives of New England may be well conjectured to be from the scattered Trojans, after such times as Brutus departed from Latium."[17] Sometimes the magnitude of their deductions caused these amateur historians to get out of breath.

A more sober picture of the Mississippi Valley antiquities can be gained from perusing the maps executed by mid-nineteenth-century land surveyors. Like most prehistoric undertakings, the scale and extent of the mounds built by the Indians' ancestors is bound to astonish even people addicted to superlatives. The number of tumuli in Ohio alone is estimated to have been ten thousand. Truncated and stepped pyramids; mounds for sacrificial, mortuary, and observation purposes; sacred enclosures and defense works—all once occupied the entire river basin where today often the only architectural landmarks are service stations and roadside diners. Earthworks spreading over as many as 600 acres attest to the builders' mastery of both geometric patterns and stylized organic forms. The great accuracy with which the squares and circles, ellipses and polygons are executed makes it seem likely that standard measures and some rudimentary instruments had been used in their execution. A curious sideline are huge earth structures, bas-reliefs in the shape of beasts, birds, and reptiles. However, no structure even vaguely resembling Stonehenge came to light on North American territory.

60. Opposite page: The greatest concentration of the mound builder's monuments occur in Ohio and Mississippi, where thousands of defense works and sacred enclosures, often up to a mile in circumference, cover the plains. The accuracy of execution implies that the builders had standards of measurements and knew how to determine angles. From Ancient Monuments of the Mississippi Valley *by E. G. Squier and E. H. Davis.*

Englishmen who, unlike Americans, do have proof that their country was colonized by the Romans, are still at pains to puzzle out the paternity of their number-one national monument. Blacket did not have the last word on Stonehenge; in 1965 another American had a go at it. In his book *Stonehenge Decoded*, the astronomer Gerald S. Hawkins added one more interpretation to the controversial pile.

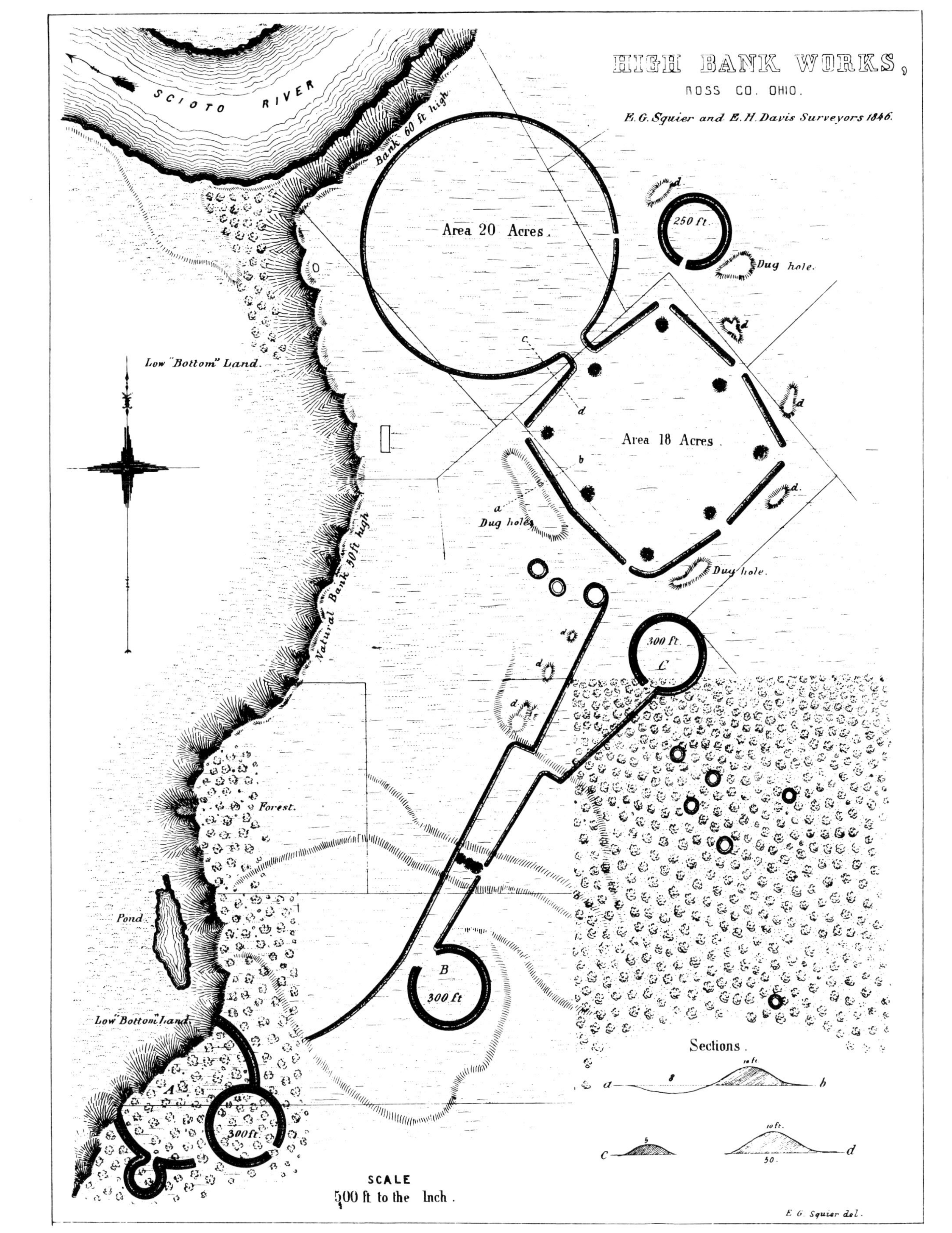
HIGH BANK WORKS,
ROSS CO. OHIO.
E. G. Squier and E. H. Davis Surveyors 1846.
SCIOTO RIVER
Bank 60 ft high
Area 20 Acres.
250 ft.
Dug hole.
Area 18 Acres.
Low "Bottom" Land.
Dug holes
Dug hole.
Natural Bank 30 ft high
300 ft.
C
Forest.
Pond.
B
300 ft
Low "Bottom" Land.
A
300 ft.
Sections.
SCALE
500 ft to the Inch.
E. G. Squier del.

Aided by the latest scientific tools, he pushed back Stonehenge's age into 2000–2200 B.C. He also was able to establish that the structure represented "a sophisticated and brilliantly conceived astronomical observatory."[18] (As mentioned above, Hawkins's thesis had been anticipated forty years earlier by Trotter's article "Stonehenge as an Astronomical Instrument.") What made these revelations so poignant is the fact that Stone Age men were indeed conversant with astronomy. Although they earned our respect as hefty masons, few people are willing to believe that they were less childish than children. However, it would seem that these preliterate people could not have been blind to celestial events, the more so as they were wont to regulate their lives by them. By worshipping sun, moon, and the stars, they took a lively interest in their paths, and may have felt a need for keeping track of them.

None of the men who fabled and surmised upon the purpose of the henges adequately answered the question of how their builders went about their task. Architecture has its measure of follies, major and minor, yet all of them are products of advanced building technologies. Such technologies do not materialize in a society of primitive herds- and husbandmen; they are characteristic of full-blown civilizations. No vestiges of such civilizations could be detected in the Salisbury plain. How, then, can one explain that a race of men who may have

61. The stones for Cuzco's fortress Sacsahuaman were transported over 40 miles through roadless mountain territory without the aid of vehicles or draught animals. They average 15 × 12 × 10 feet or about the size of an American bedroom, while the biggest of them measures 27 × 14 × 12 feet. (Courtesy American Museum of Natural History)

barely managed to keep body and soul together fell victim to an urge so strong as to move the most unwieldy rocks over land and sea for hundreds of miles, to set them up in a place destitute of charm? No lake or river extends an invitation to tarry; no hill or mesa asks to be crowned with a man-made pinnacle. The persistence with which men played their strenuous games of shifting and laying boulder-size stones leaves one perplexed; transporting the jeroboams is in itself a science, now lost. Only intense aversion to littleness could account for the compulsion to undertake superhuman feats.

One might as well attribute this faculty for juggling rocks of several hundred tons of weight to the aid of some higher agency. Or perhaps these vanished races possessed the secret of levitation. The Spanish historian Garcilaso de la Vega thought, apropos the Peruvian fortress of Sacsahuaman, that "those who have seen it, and studied it with attention, will be let not alone to imagine, but to believe, that it was reared by enchantment—by demons, and not by men, because of the number and size of the stones placed in the three walls, which are rather cliffs than walls, and which it is impossible to believe were cut out of quarries, since the Indians had neither iron nor steel wherewith to extract and shape them." Not only are we unable to duplicate the early builders' feats, we still haven't discovered how they were accomplished. (The largest stone at Sacsahuaman has a computed weight of

62. A closeup of Sacsahuaman's triple rampart reveals the masonry's intricacy. Some blocks are shaped with re-entrant angles to lock with each other; all of them were fitted into place without cranes or pulleys. The two figures at the foot of the wall give an idea of the stones' size.

361 tons.) "It passes the power of imagination," Garcilaso noted, "to conceive how so many and great stones could be so accurately fitted together as scarcely to admit the insertion of the point of a knife between them. Many are indeed so well fitted that the joint can hardly be discovered."[19]

The ultimate reason for the monumental enterprises remains obscure. The Peruvian fortress vouchsafed security, but none of the henges, circles and alleys that dot England, could have been counted among life's necessities. They represent, it would seem, a luxury that primitive people could ill afford. But then, functionalism was not an overriding consideration in those distant times. All human efforts may have been expended in nonutilitarian projects. In fact, to assume that functionalism stood at the cradle of architecture merely betrays a materialistic orientation. In the olden days sheer muscle power seems to have been the chief propellant for communal building activities. The monstrous exertions probably served as outlet for the pent-up energy that less intelligent races expend on warfare. Whether men joyously lent a helping hand or slaved under a scourge the results do not tell. No computer will uncover these secrets; only the acuity of the human mind may be able to wrest them from the stones. We simply have to make do without those human-interest stories that lend sparkle to the most arid matter. Until we learn more about the shadowy races that cultivated the art of juggling boulders in preference to playing some aboriginal cricket, we are free to conjecture about the substance of their life and to try as well as we can to dissociate their architectural standards from ours. (What would the henge builders have thought of the latter-day British architect's recipe for instant stone masonry—a brick wall "to be dashed with quick lime and sharp sand, well mixed together, and colored so as to imitate stone."?[20])

It stands to reason that the urge to build dwellings proper arose quite late. The need for some solid kind of shelter was felt only after enough human energy had been expended in the erection of extravagancies. The humble house did not precede monuments but followed them.

Analogies with apparel come to mind: The assumption that garments were devised to clothe the human body (more precisely, its various orifices and appendages), self-evident as it may appear to people living in a dour climate—dour in both the meteorological and the mental sense—has proved to be untenable. In the friendly zones the stress is, more often than not, on nonutilitarian, playful coverings. In other words, *l'art pour l'art* and *l'architecture pour l'architecture* take precedence by seniority over products with a practical purpose. The numerical superiority of both so-called functional architecture and functional clothes can be no doubt attributed to that puritanical streak which always surfaces in unstable societies.

Eight years after his first book appeared, Hawkins published a sequel, *Beyond Stonehenge*, that strengthened his hypotheses by delineating a global panorama of architecture in the service of astronomy. The pyramids and temples from Egypt to South America come in for his scrutiny, and neither are the North American mound builders' feats forgotten. Actually, Hawkins's revelation was long in coming; anybody who stood at the top of Delhi's Samrat Yantra, one of the world's two largest sundials, and looked (as I did) toward the

63. Opposite page: The gigantic astronomical instruments at Delhi, India, built in 1724, are among the most imposing examples of abstract architecture. One of the two Ram Yantras, cylindrical instruments for reading the sun's altitude, can be seen in the background. Had Stukeley known it, he probably would have guessed that Stonehenge (near right) was an observatory.

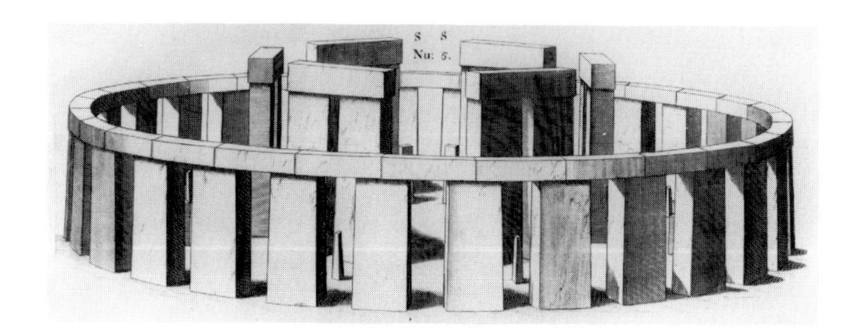

nearby circular astronomical instruments called Ram Yantras, may have been reminded, if not of Stonehenge proper, at least of its mildly fanciful reconstruction by Inigo Jones. To be sure, the Delhi roundhouses have three stories, but the number of the pillars—thirty —is the same.

The Indian observatories are in an architectural category all of their own. The generic classification may be misleading if one has the Mount Wilson type in mind because here, the instruments are not protected by a dome but simply left open to the sky. Sculpture gardens would be a more fitting designation for them. Yet they also have a certain affinity with the megalithic monuments, indeed, more so than with pedigreed architecture. They have no doors, no windows, and no roof; their concept and execution are uncompromisingly abstract.

64. A maharaja's astronomical toyland. Jaipur, India.

They appeared rather late on the architectural scene, being contemporary with Philadelphia's Independence Hall. There was no need to call upon an architect to draw up their plans and elevations; they are the brainchildren of the eighteenth-century ruler of Jaipur State, Maharaja Jai Singh II, city planner and aficionado of astronomy. (He built four more observatories in Jaipur, Benares, Ujjain, and Muttra.) Desirous to obtain precise tools for examining the motions of celestial bodies, he had the fragile brass instruments of his day enlarged a hundredfold and translated into masonry. Although the scientific harvest did not come up to his expectations, the tangible results are nevertheless astonishing—an Indian rococo scenery measured in languorous rhythms: curious shell-like depressions; twisted and broken curves accompanied by the narrowest of steps whose very sight induces dizziness.

Jai Singh died in 1743, two hundred years after Copernicus. As one chronicler remarked, "his wives, concubines, and *science* expired with him on his funeral pyre."[21] The maharaja may take credit for having designed the Samrat Yantra and Jai Prakas, but the observatories' basic idea was not his. Arabian and Persian astronomers had built gigantic instruments before him.

65. Opposite page: Interior of a Ram Yantra.

66. This detail from Delhi's Samrat Yantra, an equinoctial sundial, shows steps winding, as it were, through petrified friction gears.

Beyond the unmarked boundaries of architecture beckon minor eccentricities. The pick of the basket is Lindenthal. The name of this Cologne suburb does not ring a bell with students of avant-garde architecture, as does, for instance, Stuttgart's Weissenhofsiedlung. Yet once there stood on its site a complex of houses that today might carry off all those medals and prizes which our arbiters of taste periodically bestow on presumably imaginative architects. Except for archaeologists interested in the culture of the *Bandkeramiker* (makers of curvilinearly decorated ceramics), this early domestic architecture remained largely unknown, perhaps because it produced no ruins. Some post holes and cavities in the ground are all that's left of the former dwellings. However, despite the meager residues, we are reasonably well informed about the dwellings's original appearance.

If the nineteenth-century discoveries of cave paintings and cave sculptures obliged us to drastically revise our notions of early man's artistic talents and aspirations, here we have proof that he also was well equipped to contrive and shape habitable space of transcendent bodily and mental comfort. In matters of aesthetics—if such a weighty word is at all applicable to the train of thought of such unselfconscious people—the Lindenthalers were a good deal more imaginative than most mankind that succeeded them.

For a neolithic village, the houses were immoderately complicated. They were neither round nor angular, as primitive houses tend to be all over the world. Instead, they were of irregular outline, and not two of them alike. They also were surprisingly large. Never at a loss for a suitable term, the excavators dubbed them *Kurvenkomplexbau*, that is, complex curvilinear structures.[22] Extraordinary by any standards, they do not strike one as Stone Age peoples' gropings; for that they are far too methodical. None of the houses' shapes seems to be accidental. The enclosing wickerwork walls, for instance, do not follow the concavities in the ground. Neither clumsiness nor sloppiness would seem to account for their erratic contours. Should one, then, attribute them to the builders' playfulness? Primitivism is a ticklish subject, the word *primitive* vexaciously ambiguous. Whether applied to religion, race, or art, it has little meaning unless appended by wordy circumlocutions. Benin sculpture, for example, has been classified as primitive perhaps because the Beni are a pure Negro tribe, yet the best of it is in no way inferior to sculpture of the High Renaissance. So-called primitive religions may be far more attractive—and more gladdening—than many of our faiths. Again, modern societies have as big a share of primitive members as have technologically backward nations.

Primitive art became legitimized long before the first Museum of Primitive Art opened its doors. With (so-called) primitive architecture we stand on less secure ground. Engrossing as it is, it does not much lend itself to comprehensive classification. Its boundaries are hard to define. The more carefully one looks at the reconstruction of a Lindenthal house, the more one hesitates to call it primitive. Besides, it is not an odd example of its kind; it is typical for the houses of the entire settlement. Its irregular plan and uneven floor are characteristic for the plans and floors of all the other houses. No convincing explanation for the wickerwork screens's twisting lines suggests itself. One would assume that the ground was too irregular to allow for a more consistent placing of the roof's vertical posts. This was not the case; apparently, the winding walls were what the inhabitants wanted. They frankly preferred organic shapes such as mesentery and kidney to geometric ones.

If we never developed a taste for irregular space configurations like the Lindenthalers' it is because we do not think of our houses in terms of a wrap but of a box. Ours are hard-edge containers with perfectly level floors, and no deep-pile carpets, no water beds will confer sensuousness upon them. Apart from some pardonable affectations, curved walls do not occur in the domestic architecture of civilizations prejudiced in favor of angularity and edginess. From a brick to a sheet of glass, from a book page to a bed, our domesticity is molded in the cast of the right angle. Except for the right-hand side of a grand piano, any departure from squareness hints at dissoluteness.

67. *Sinuous contours, meandering space, and cup-shaped lolling pads point to a calculated voluptuousness in the dwellings at the neolithic settlement of Köln-Lindenthal in Germany. Its vestiges upset any notions that all early houses were plain and disconsolately austere. Cooking being done outdoors, the earth-sculpted living room floor was the family playground. House no. 36 from* Die bandkeramische Ansiedlung bei Köln-Lindenthal *by Werner Buttler and Waldemar Haberey.*

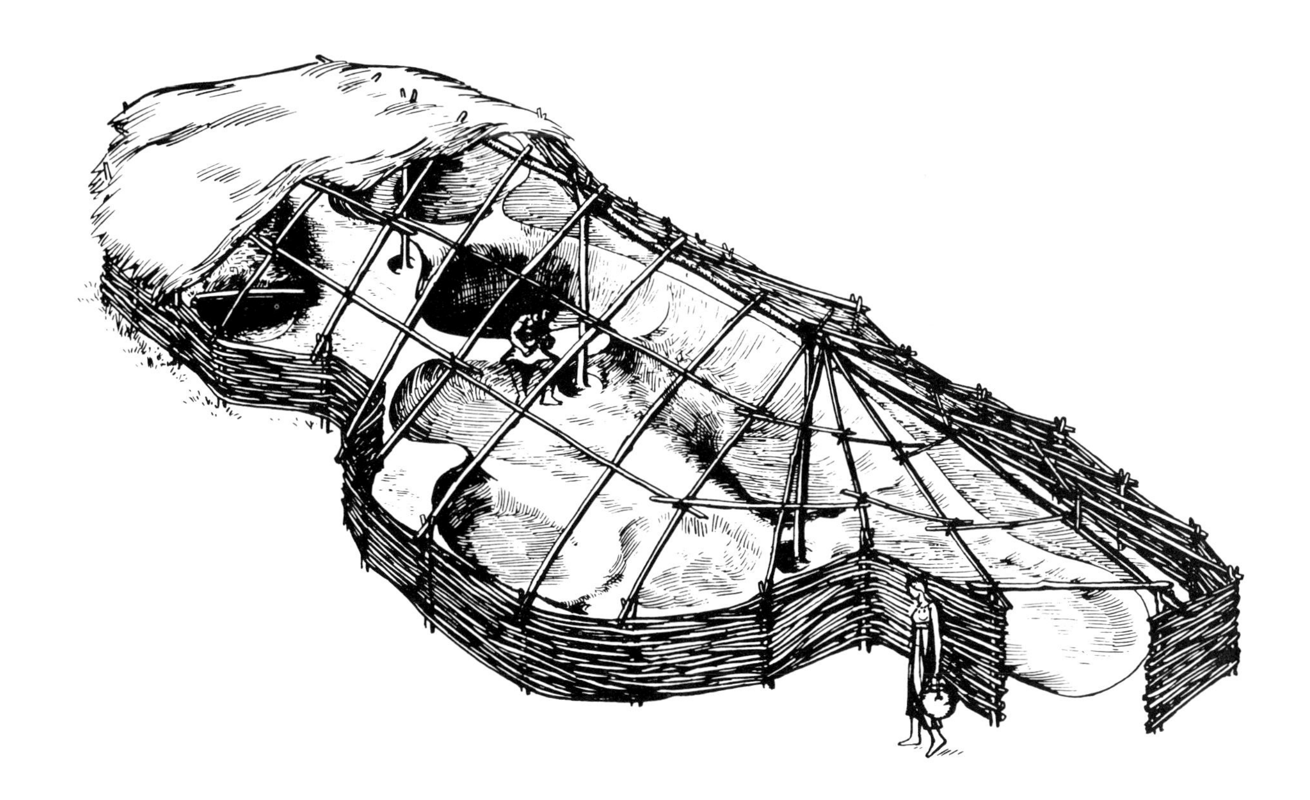

What lifts the Lindenthal houses above the commonplace is their sculpted floor. Rarely has pleasure in free-wheeling form been as felicitously expressed as in the shell-shaped cavities that perfuse their interiors. Pausing for a moment in the amble of our flat-footed way of life, and looking back through the mists of time to the *Bandkeramikers'* lairs, it seems that somewhere in the course of human advancement we lost our capacity for enjoying well-turned-out space. What probably happened is that our *Fingerspitzengefühl*—our "fingertip feeling"—progressively deserted us while numbness crept up on our toes and buttocks, body parts traditionally hidden. Yet it is exactly these parts that are instrumental in probing the ground on which we walk, stand, or sit, for they, like no others, keep us in close touch with our surroundings. Custom, however, demands that our posterior be supported by a chair, at least a dozen inches above the floor, while our feet be encased in, preferably, hard-shell shoes.

The Lindenthalers got one better of both primitive and civilized savages by scooping out of their houses' floor a sequence of hollows—inverted hillocks, as it were—that invite one to loll, to insinuate oneself into the soil, so to speak, albeit soil of a highly stylized kind. "A man, an animal, an almond, all find maximum repose in a shell," says the forementioned Bachelard, the author of *The Poetics of Space*; "to curl up belongs to the phenomenology of the verb to inhabit, and only those who have learned to do so can inhabit with intensity."[23] Clearly, the floor of these houses is admirably suited for curling up. It is pure sculpture, to be felt not just with one's fingertips but by the deft touch of one's whole body.

The assumption that early man in search of shelter came upon natural hollows in the ground which he enlarged to family size does not hold good at Lindenthal. There, the terrain is flat. The floors' relief is man-made and does not take advantage of topographical accidents. The archaeologists who unearthed the settlement were unable to account for what they call the "intended irregularity" of house plan and floor surface; "a plausible explanation," they wrote, "why the neolithic people of the Bandkeramiker built such peculiar dwellings is still missing."[24] The least implausible explanation would be to attribute to them a rare hedonistic apprehension of the poetics of space.

One cannot help noticing that the perspective drawing on page 103, lovingly rendered though it is, misses the charm of the actual premises by conveying an impression of unmitigated austerity. The interiors were probably never as stark and dark as the reconstruction implies. Several overhead openings, or a few strategically placed translucent animal hides, stretched over purlins, may have let in some light. There was no lack of tools and pots that today would have a

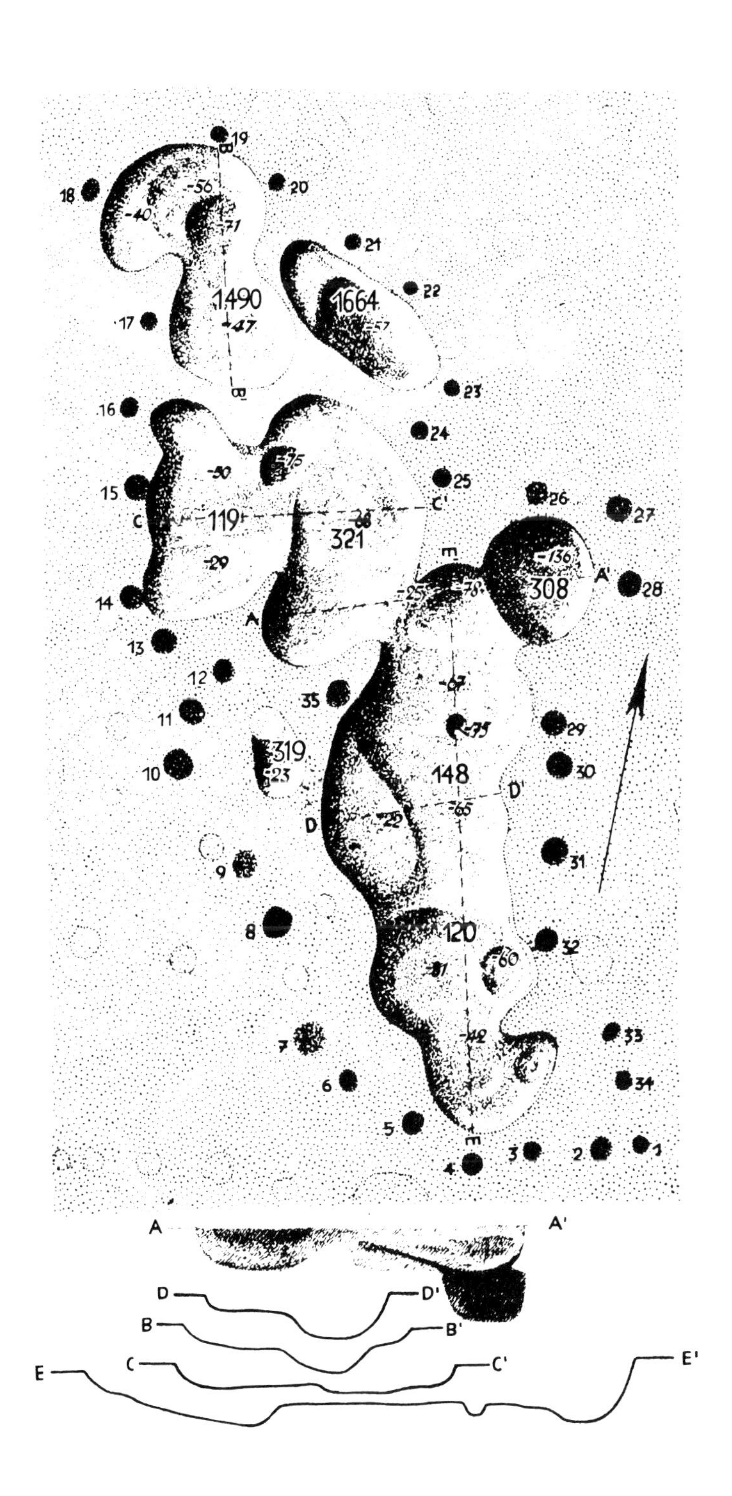

68. *The plan of house no. 30 at Köln-Lindenthal illustrates the free flow of walls and floor. Numbered dots indicate wooden uprights; the hollows in the ground, which average 30 inches in depth, are an ingenious solution for providing built-in sitting and sleeping accommodations without encumbering interior space with bulky furniture. (For a comparison with neolithic seating comfort, see the prehistoric sofa and chair in figs. 262 and 263.) From* Die bandkeramische Ansiedlung bei Köln-Lindenthal *by* Werner Buttler *and* Waldemar Haberey.

69. Relief plan of a prehistoric house at Lissdorf, Thuringia, which clearly shows the floor's convexities. The snug dents and dips scooped out of the ground range from man-size to family-size. (After Schuchhardt)

place in museums, nor of baskets, now crumbled into dust, while some makeshift upholstery—leaves, mosses, and animal skins—served as padding for the scalloped indentations in the ground. These ghosts of beds and sofas, melting into the floor, formed, as it were, landscaped furniture. Surely, this was shelter not for stoics but for voluptuaries.

The irregular pits deserve further comment. Pits of a kind enjoyed a passing vogue in the mid–twentieth century after Christ. Houseowners, itching for household extravagancies, had a portion of their living room floor deepened by several feet to form a lower level, a negative estrade, so to say. This sunken space—a sort of square womb for faltering souls—was bordered on several sides by upholstered seats which provided the requisite coziness. (The closest approximation to the Lindenthal house's floor was an "uneven floor," a side show in my 1944 exhibition "Are Clothes Modern?" at New York's Museum of Modern Art. Constructed from wood and plaster, and covered with paper that imitated marble, it was meant to simulate a stone relief of irregular, undulating contours. This glimpse of a more adventurous flooring than the kind we are used to did not open up new avenues of domestic luxury. An uneven floor was inimical to our footwear, while going barefoot at home at that time smacked of exhibitionism or worse.)

To return to that formidable cultural backlog, megalithic structures, hallowed, magical, or otherwise: Man's urge to put up stones, especially stone circles, whether for rituals or as places of assembly—Avebury could easily seat two hundred fifty thousand people—was widespread. It is shared, as we have seen, by the infant ape, a tendency that escaped Darwin, just when its tidings might have been helpful to shore up some of his arguments. So far nobody bothered to investigate the sort of satisfaction derived from being enclosed by the sketchiest of circular structures. Do man and ape put up the stones because they find them pleasant to look at? Do they consciously seek their abstract embrace? Does the act of building circles bring pleasure? Moreover, how much are a builder's feelings colored by anticipation or recollection? Stone Age man left no clues, and the chimpanzee, acting from his inner promptings, is unable to take us into his confidence.

One major hindrance in architecturally legitimizing the megalithic monuments—admittedly an exceedingly controversial subject—lies in drawing the line between sculpture and architecture. Whereas everybody can tell the difference between a painting and a piece of sculpture, the distinction between sculpture and architecture is less marked. The problem that architecture poses itself is, by definition,

70. By erecting the first wall, man arrived at a point in his evolution that was as sharply defined as when he got up from all fours and stood on his legs. Building the first wall, he became, mentally, a biped. The illustration shows a place of assembly of the prehistoric inhabitants of the Canary island of Hierro. From Berthelot's Antiquités canariennes *(1879).*

to solve how best to enclose space for human occupancy. The Pyramids, a classical borderline case, certainly do not fit this concept, and yet they figure prominently in architectural history.

Conversely, there are instances where buildings mimic sculpture. The Washington Monument—to choose a well-known example—is not an obelisk but a tower that was, with interruptions, twenty-nine years abuilding. True obelisks are monoliths; they are anything but native to the Western world. All the famous obelisks that stand in Rome—twelve of them, to be exact—in Paris, London, and New York are loot from Egypt. In their homeland they were not put up as decorative accents; they fulfilled a symbolic function, much like the cross in our part of the world. Yet whereas the cross stands for torture and death, the Egyptian obelisks represent joyous exclamation marks, pillars dedicated to the sun gods. What makes them relevant to our discussion is that they are streamlined versions of Stone Age menhirs. A note on the terminology of megaliths will help to get them into focus.

Menhirs, together with dolmens and cromlechs, constitute the ABC of sculpture-architecture. *Menhirs* are single upright stones of considerable bulk, either dressed or in their raw state. Their habitat extends from the Atlantic coast of France all the way to Siberia and China, from Scandinavia to Africa. They may or may not commemorate an event, mark a grave, or be intended as phallic or anthropomorphic representations—one guess is as good as the other. In some instances, they were probably connected with fertility cults or served as altars for human or animal sacrifice. The results are amply rewarding: triumphant if uncouth columns, the tallest of them a dozen times the height of the tallest man. The giant menhir Men-er-Hroec'h at Locmariaquer in Brittany, now lying on the ground in four pieces, measures 67½ feet, only 18 inches less than Cleopatra's Needle, the monolithic obelisk from Heliopolis in New York's Central Park.

Cromlechs are menhirs that form a circle—in France, that is. In Britain *cromlechs* are synonymous with *dolmens*. A cromlech, straightened out, becomes an *alignment*. This is the archaeologists' designation for the rows of upright stones scattered along 15 miles of Brittany's southern coast. It would be hard to find a more prosaic name for them, but then they have never been acknowledged as works of art. Pedants may point out that the stones do not even align. Rather they are wandering, Indian file, through moors and fields, like soldiers tired after a day's march. The long lines wiggle, and some stones seem to be at the point of breaking ranks. But then, aligning them was probably the last thing in the mind of the men who put them up, since straight lines, much as right angles and symmetry—

71. *Poetically licensed artists do not feel obliged to stick to the truth. This nineteenth-century version of Carnac's battalions of menhirs conjures up a one-time good order that never was. The engraver has stretched the stones and serried their ranks.*

the highest attributes of grand architecture—were still in the future. The most impressive concentration of these stones is near the village of Carnac on the Morbihan peninsula, not to be confused with Karnak, a village in Egypt, equally famous for its architectural antiquities. The rows of Carnac's 2,935 menhirs stretch for 2½ miles. Thousands of years of rain and hail have diminished their size; loot for building material reduced their number to a fraction of the fifteen thousand stones counted in the sixteenth century. Yet even in their dilapidated state they are a powerful challenge to one's imagination.

Until a hundred years ago they were thought to be remnants of an encampment of Caesar's troops who used the monoliths for securing their tents against the furious winds that lash the coast. Another fanciful surmise identified them as pillars of a dracontium,[25] a temple dedicated to the cult of serpents. More sober interpreters saw in them "social memorials set up along the route from a settlement to a boat landing."[26] Still others maintained that they are cemeteries. Today, they are believed to be relics of an Asian civilization established some three to four thousand years ago by people prospecting for deposits of tin, a metal much in demand prior to the dawn of history.[27]

The question whether the monoliths form architectural spaces or, generally speaking, whether rows of free-standing roofless pillars add up to an ever so sketchy, indeed, illusory enclosure cannot be answered to everybody's satisfaction. Analogies with temple ruins are not helpful because temple columns are rigidly tailored to geometrical formulas whereas the Breton menhirs are at ease. Not ruins but an unfinished building on a grandiose scale presents architectural space comparable to Carnac—the mosque that Jakub-el-Mansur, the twelfth-century Moorish prince, intended for Rabat. Vast enough for a whole army to worship in simultaneously, it never progressed beyond the preliminaries, except for a magnificent minaret. In its present state, it amounts to no more than a field, seeded, as it were, with identical columns in closed ranks and with the canopy of Heaven for a roof.

72. Yakub-el-Mansur's uncompleted mosque at Salé, Morocco.

Carnac's stones are not without solemnity. At Le Menec they form a kind of open-air cathedral of eleven naves a quarter-mile long. The Bretons never prayed in it; they used it, instead, as a ballroom for their rustic dance festivals. But then, once upon a time, dancing was a form of praying and surely more wholesome than kneeling. Ordinarily we don't perform ritual dances at the shrine of a deity or a saint. Yet many a traveler in Spain is surprised to see a ballet performed in front of the high altar in Seville's cathedral by boys in seventeenth-century costume, with plumed hats and castanets, going through their gyrations to the accompaniment of a chamber orchestra. Although the Bretons did not venerate the monoliths, they attributed magic powers to them. Women took scrapings from the stones as a sympathetic remedy for barrenness.

The step from the lone monolith to the trilithon, or what we call jambs and lintel, led man to the threshold of architectural awareness.

73. Trilithon at the Maltese sanctuary of Mnaijdra. The pockmarks represent a cosmetic sort of stone ornament.

It marks the transition from sculpture to architecture. When the three stones are substantial enough, they form the simplest of dolmen. As a rule, however, a *dolmen* consists of a number of upright stones, or slabs, forming parallel walls that are topped by one or several flat outsize stones. The largest ones may weigh as much as 40 tons. Dolmens are encountered chiefly in Europe but also in Asia as far east as the Japanese islands. A tiny country like Denmark has some 4,700 of them.

Except for the Blackets and Brackenridges, most antiquarians and scholars are reluctant to forge any links between the prehistoric builders of the various continents. Yet there is nothing frivolous about conjecturing the existence of early intercultural relations—traveling tastemakers, pre-Homeric gazetteers, old geezers with plenty of muscle left for pollinating the flowering youth with their architectural wisdom. Either news traveled far or man's building urge manifests

74. Dolmen near Puenteareas in the province of Pontevedra, Spain.

itself in fairly predictable patterns.

Cromlechs and dolmens presumably were once covered by earth mounds that have long been carried off by wind and rain so that what we see today are the mere skeletons of the former structures. But on this point opinions differ. At any rate, in the dolmens we can discern one of the oldest models of the house. Drafty it may have been, but it also was near-indestructible. Not that neolithic man and his kin intended to inhabit it while they were still alive; only the dead qualified for it. We thus meet already in the Stone Age with the sort of Victorian mentality which demanded that a genteel household have a parlor—not for the convenience of the inhabitants but for visitors, wakes, and similar melancholy occasions.

When Christian saints relieved the pagan gods of their official position, the time had come for menhirs and dolmens to conform to the new faith. To speed up matters, the Council of Arles (A.D. 452) decreed that stone worshippers who neglected to forswear and destroy

75. The dolmen "La Roche aux Fées" in the French department of Ille-et-Vilaine.

them should be found guilty of sacrilege. Luckily, the clergy was not averse to doing some missionary work on the side. "Take holy water," advised Pope Gregor the Great, "and sprinkle it on these shrines, build altars and place relics in them. For if the shrines are well built, it is essential that they should be changed from the worship of devils to the service of God."[28] (There is irony in the fact that the very method which His Holiness prescribed for the rehabilitation of the offensive monuments was identical with heathen practices.) At any rate, many menhirs were spared an ignoble fate by being blessed and sprinkled, which was all that was needed to accomplish their conversion. In France, ever a citadel of the true faith, a finishing touch was added by planting a prophylactic cross on the megalith.

Curiously, the question of sex—that is, of the stones' sex—seems to have eluded both clergy and laity, and I don't mean the obvious phallic implications. Admittedly, some of the stone piles are less than awe-inspiring, amounting to no more than ponderous pranks. However, anybody even only slightly acquainted with the esoteric teachings of the Orient knows that stones, like beasts, are male and female. Any respectable Japanese gardener will oblige with a discourse on the subject. The people who put up the thousands of shapely boulders on the shores of the Atlantic had come from the other end of the Eurasian continent and probably were versed in stone lore. They also may have brought with them poetic notions about, and a love for, stones, which they expressed in their gigantic compositions.

76. In India, on a high plateau in the Dekkan, stand more than two thousand dolmens that are noteworthy for a feature rarely found elsewhere—a small opening in one of the lateral stones, made, it has been assumed, for serving food to the dead or as a sort of escape hatch for their souls. From Pre-Historic Times *by John Lubbock.*

A variant of these granite houses of cards is the *taula*, characteristic of Minorca, the Cinderella of the Balearic Islands. Better known as the birthplace of mayonnaise than for its antiquities, Minorca still preserves the air of mystery one associates with the billets of such distinguished travelers as Aeneas and Odysseus. The island is an archaeological paradise. Hundreds of megalithic monuments stand in the pastures, hidden from sight by *cercas*, tall stone walls that divide the small peasant holdings. Not to miss them, it is advisable to go looking for them on horseback.

Since Minorca's monuments enjoy a certain anonymity, they have

77. Taula, a Stone Age version of column and capital, incomparably more forceful than its descendants, the classical orders. Minorca, Spain.

remained undisturbed so far. No directional signs denote their whereabouts, no explanatory tablets or warnings mar the surroundings. Nor does one come across empty bottles and discarded film wrappers, the offal of the tourist trade. Such small matters should not be dismissed as unimportant. Today, when packaged travelers descend on ruins like flies on carrion in the vain hope to assuage their boredom, it may soon be more prudent to restrict one's aesthetic pilgrimages to contemplating the pages of coffee table books.

A well-preserved taula impresses one as something right out of the studio of an artist who embraced the tenets of uncompromising simplicity. It is minimal sculpture-architecture on a noble scale: an upright slab topped by a horizontal, tablelike one. In fact, taula means *tabula*, or table. In contrast to menhirs and dolmens, here the stones have been expertly dressed, with sharp edges and smooth surfaces. The taulas' function has not been unequivocally established. Some think that they were tombstones. Others see in them central supports for a roof that disappeared long ago. Whatever the correct interpretation, the spare volumes, beautifully set off by the landscape, are among the most eloquent testimonials to man's creative talents.

Another archaic type of structure, encountered only on Minorca, are two dozen *naus*, or navetas, oblong pyramids with rounded corners, in the form of an upturned boat. Anybody who thinks that this extravagant shape is totally unrelated to architecture as we understand it ought to note that a couple of years ago Kenzo Tange, the peerless Japanese architect, built—in concrete—a gymnasium in the form of a boat, complete with nautical features, except oars. A *jeu d'esprit* rather than an epoch-making work of art, it nevertheless found its place in the annals of modern architecture.

Navetas are fossil bones of contention, and opinions about their erstwhile destination are divided. Believed to be collective graves—more precisely, pantheons of some unknown soldier—they may just as well be the petrified version of a coffin in the form of a canoe in which a corpse is set out on his journey to the Land of the Dead, the water burial of Oceania.

With *talayots*, the third Balearic specialty, we are getting closer to human habitations. Talayots are truncated cones of cyclopean masonry, either of a round or squarish plan, as much as 100 feet in diameter and about 40 feet high. They usually contain a single chamber, reached by an entrance high above ground. The walls are substantial enough to accommodate spiral ramps in their thickness, a feature also known from Brochs, the towers of the Picts. Originally built as tombs, they served as guardhouses from remotest times—their name means watchtower in Catalan. There are over one thousand of them on Majorca, less than three hundred on Minorca.

These three pictures exemplify various stages of centrally supported structures.

78. Right: Conjectural section of a stone hut, about 15 feet in diameter, at Hayes Common, Kent. An uprooted tree supports the roof. From The Neolithic Dwelling *by George Clinch.*

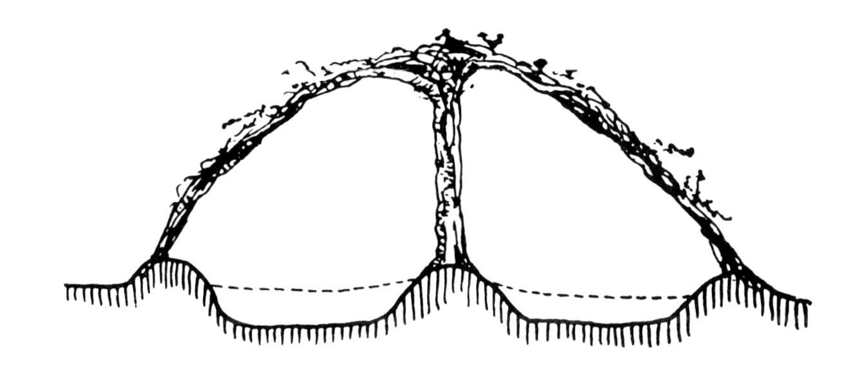

79. Right: Interior of a reconstructed forge from Mšecké Žehrovice, Czechoslovakia. Irregular, recessed floor; thatched roof with a central support. Latène period, second century B.C. *Outdoor museum at Asparn/ Zaya, Austria.*

80. Opposite page: A solidified stage is represented by this Etruscan tomb, reconstructed from excavated material. The pillar was indispensable for propping up the pseudocupola, being weighed down by the mass of earth. Archaeological Museum, Florence.

81. Nau d'Es tudons at Ciudadela on the Balearic island of Minorca is one of the two dozen prehistoric collective tombs called navetas, *dating from the twentieth to the fifteenth century* B.C. *The elongated structure encloses a rectangular chamber.*

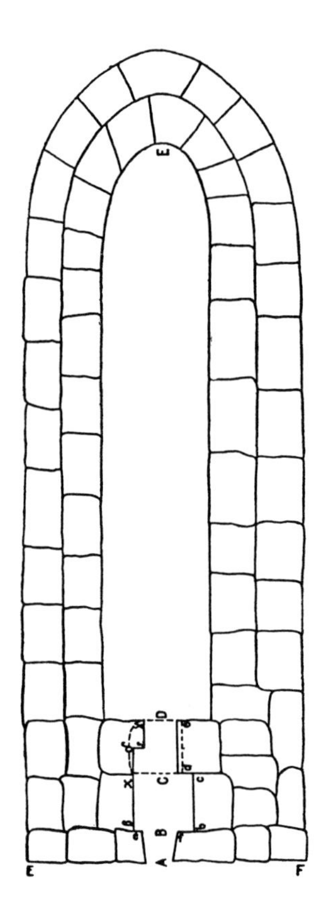

82. Plan of the Nau d'Es tudons on Minorca. (After Bezzenberger)

(*Webster*'s misspelling of *Mallorca* curdles every literate Spaniard's blood. Alas, this accords with Rudofsky's Law, which states, "Every foreign word correctly misspelled, mispronounced, or misapplied becomes automatically an American word.") In times of distress talayots provided refuge, but their alleged use as fortresses probably was limited. Although Majorca was invaded at some time or other by almost every Mediterranean nation, there are no records of war.

"The second millennium," writes N. K. Sandars, "was a time of great building activity in the central and western Mediterranean, and towards its end in the early first millennium the impetus moved away from monumental tomb buildings to secular work: towers, castles, villages, and even little towns. The flower of all this activity was in Sardinia."[29] In those times Sardinia possessed a military architecture, a network of superbly constructed round stone towers that either stood out as solitary sentinels or lorded it over a village. They are called *nuraghi*, probably a dialectic mangling of *muraglie*, walls. Their great number—about three to four thousand by Baedeker's count, sixty-five hundred according to Sandars, seven thousand if you believe the *Encyclopaedia of World Art*—suggests that Sardinia had a stormy prehistoric past. The enemy may very well have been in the midst of the people, a society half-pastoral, half-martial. Although the biggest of the nuraghi were spacious enough to admit a hundred or more persons, they were not houses for the plebs but strongholds for the aristocracy. They towered above clusters of solid stone huts which were likewise built on a round plan.

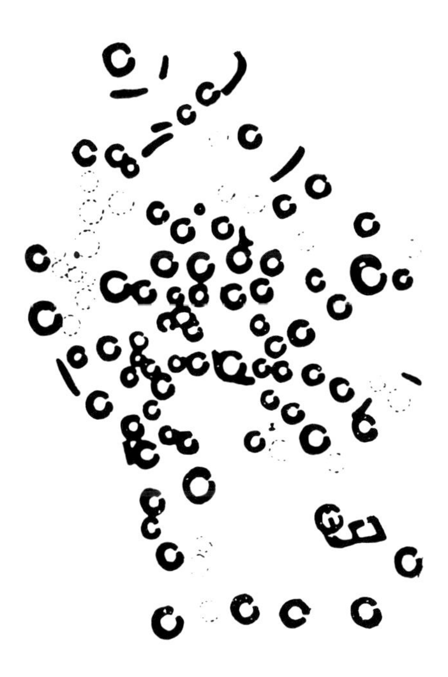

83. *Nuraghic village at Barumini, Sardinia (After Sanfilippo)*

84. *Gonessa in southwestern Sardinia is one of fifty nuraghic villages of round stone huts, from 20 to 50 feet in diameter, that once were covered with conical roofs. Vestiges of altars in the dwellings testify to an advanced state of civilization.*

85. As if nature was out to mock man's architectural conceits, she playfully put up these perplexingly realistic imitations of Stone Age buildings. What looks like a nuraghic citadel are basalt formations at the foot of Mount Etna. From Houël's Voyage pittoresque en Sicilie . . . . *(Courtesy British Museum)*

86. The grand nuragh Ortu in the Sardinian province of Iglesias, a cluster of towers in the form of truncated cones, is 65 feet high, its walls 20 feet thick. The windowless rooms suggest burial chambers rather than human habitations. Chipiez's reconstruction of the nuragh confers upon it a strong flavor of French medievalism. From Histoire de l'art dans l'antiquité *by Perrot and Chipiez.*

"Cyclopism is not simply a technique common to the whole of the Mediterranean area," says the Spanish scholar Julio Martinez Santa-Olalla, "it represents a culture which was originally confined to about the twelfth century B.C., having a wide distribution from Anatolia to the Balearics."[30] Nuraghi and talayots are paralleled by similar structures in Spain and Portugal. We have here something like the first international style in southern architecture.

87. *A glimpse into the inner court of a nuragh discloses stone masonry characteristic of prehistoric Mediterranean architecture.*

There are significant exceptions. From the Balearic Islands to Malta it is only a few hundred miles, yet the gap between the archaic architecture of the two could not be greater. Apart from menhirs, dolmens, and towers resembling those in Sardinia, Malta has some unique megalithic structures: fifteen sanctuaries, variously referred to as temples whose origin is supposed to be in collective rock-cut tombs. Cautious guesses place them at about 4000 to 5000 B.C. They do not live up to the image that we connect with the word *temple*—the classical shoebox, fenced in by colonnades. Their plan rather reminds one of baroque buildings, with elliptical rooms strung up along an axis that is sometimes bent. Enormously attractive as the ruins appear today with their stone panels the texture and color of raw silk, now running straight, now curving into an apse, the thought of how the temples looked in their prime is disconcerting. For instead of providing them with a correspondingly intricate roofscape of cupolas, half-cupolas, barrel or beehive vaults, the ur-Maltese covered them with an artificial hill.

Was it modesty that made them hide their light under a bushel, or did they not know any better? Had prehistoric notions or architecture hardened to a point where not just dolmens and taulas but also elaborately composed sanctuaries had to conform, outwardly, to a barrow, that is, to an ever so artificial cave? Of course, a lot of primitive architecture runs counter to the belief that a building's outside ought to express its inner organization. Apparently, the untutored builders were far from entertaining such a narrow concept.

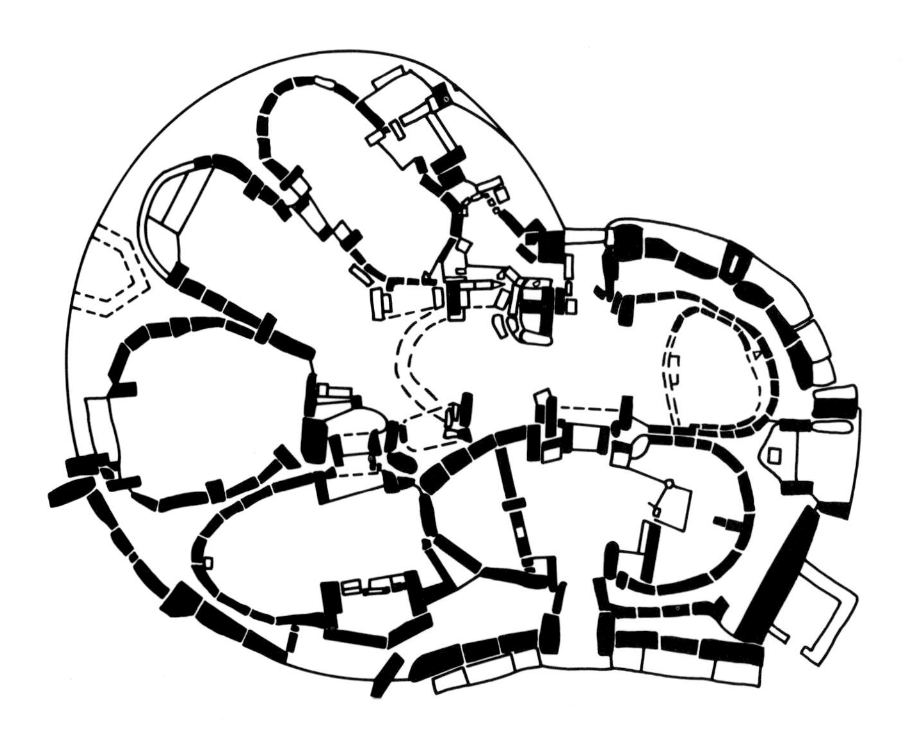

88. *Malta's neolithic villagers built their temples much like barrows, that is, covered by a large mound. Chambers were clustered informally. Plan of Hagar Qim.*

The men who composed the curvaceous spaces seem to have had an eye for obese women. The female images found in the ruins are quite unlike the idols of neighbor cultures. Of the clothed, not to say overdressed figures, a good example is the decorous "Sleeping Woman" resting on a couch (see fig. 91, p. 124). We have no clue as to whether she is a fertility goddess caught napping, or a pilgrim who has come to seek oracular revelations in trance. Maybe she is merely the votive offering of a bedridden housewife. At any rate, as women go, she is a mountain of flesh. Her small head looks even smaller against her formidable bicepses, globular breasts, and arching buttocks. (This type of the earth mother, pregnant or unrealized, that is worshipped in benign climates as the all-powerful feminine principle, reappears much later rejuvenated, slimmed, and purged of every soupçon of sensuality, as the Holy Virgin.)

89. *Although elegance is not usually associated with neolithic architecture, it exudes from every stone of Hal Tarxien, a sanctuary from Malta's Belle Epoque. The ubiquitous spiral probably was a symbol of eternity.*

90. Several thousand years before the Church adopted the apsis as focus for its places of worship, it was the dominant space in Malta's prehistoric shrines. The magnificent walls enclose an apse of Hal Tarxien.

The fragment of another buxom female, in situ at the temples of Hal Tarxien, wears the same pleated skirt over what looks like outsize bloomers. Or are the masses ballooning over her ankles part of her anatomy? Be that as it may, it would seem that this bias in favor of bulging forms—that "love of convexity"—is reflected in the layout of the sanctuaries. Double apses, linked by narrow passages, echo, diagrammatically, prominent female features—elephantine limbs topped by a miniscule head. The shapes repeat, as it were, throughout all the

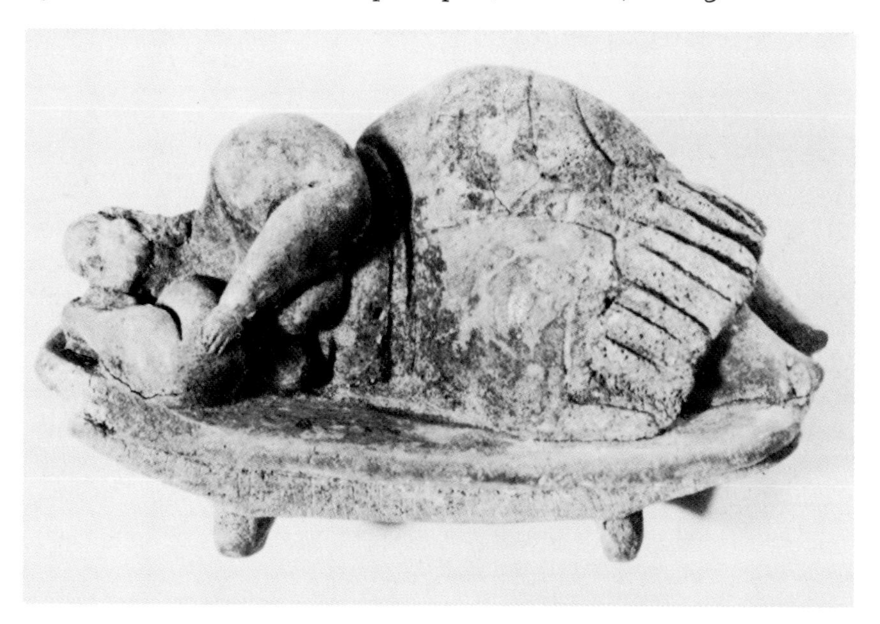

91. "Sleeping Woman," clay model (see preceding page). National Museum, Malta.

shrines that were built aboveground. Could it be that their plan concurs with the stylized anatomical outline of a recumbent female, just as the plan of many of our churches conforms to the outline of a cross? The surmise is attractive, but facts point to a different interpretation. Apsidal rooms, it seems, were nothing unusual in early times. Conspicuously absent in our houses—what could one possibly do with a semicircular space if one has no need for a house altar—domestic apses were equally at home in the houses of rustic ancient Greece, in the hamlets of the Iberian peninsula, and in huts in Denmark. Apses and *apsidiole*, universally favored in our ecclesiastical buildings, were common fare in many prehistoric types of buildings.

In 1902 a chance discovery brought to light a unique example of Maltese Stone Age architecture. While digging a well for a housing project on the outskirts of the capital, Valetta, workmen came upon what turned out to be a major archaeological find, a shrine contemporaneous with the island's temples. It might very well have remained unknown for another couple of thousand years had not the man on whose property it was found offered it for sale. This marvel, known as the Hypogeum of Hal Saflieni—neolithic monuments are not as ardently touted as classical ones—is situated, as the name implies, underground. It represents nothing less than the first full orchestration of architecture, so to speak, five thousand years before our time.

There is a pleasant robustness about it, for it was hollowed out of solid rock. It consists of a number of artificial caves, connected by ramps and stairs, and distributed on three levels, the lowest of which is 40 feet below ground. Yet there is nothing crude about it. Its spaces flow freely into each other without forming floors, walls, and ceilings in the conventional sense. The labyrinthine sequence of chambers that at different times were used for different purposes has been variously identified as a burial place, an oracle's room (which antedates those of Delphi and Cuma), treasury rooms, granaries, and a cistern. The oracle room is shaped in a way not merely to carry sound but to amplify the priest's or priestess's voice to a roar, thus stifling all doubts in the devotees' minds that they heard the goddess's own voice. A semicircular hall, its walls sculpted in imitation of pillars, lintels, and a corbeled roof, was identified as the holy of holies. Seven thousand human skeletons, unearthed in one of the chambers, suggest a vast congregation and add just the right touch of the macabre to this versatile piece of architecture. The proximity of baroque churches—pastiches of marble and porphyry, with helpings of lapis lazuli and gold leaf—make for piquant comparisons and leave one wondering which is the divine, which the profane.

How the megalithic builders arrived at their sophisticated space conceptions without the aid of inspirational prototypes of ruins is a

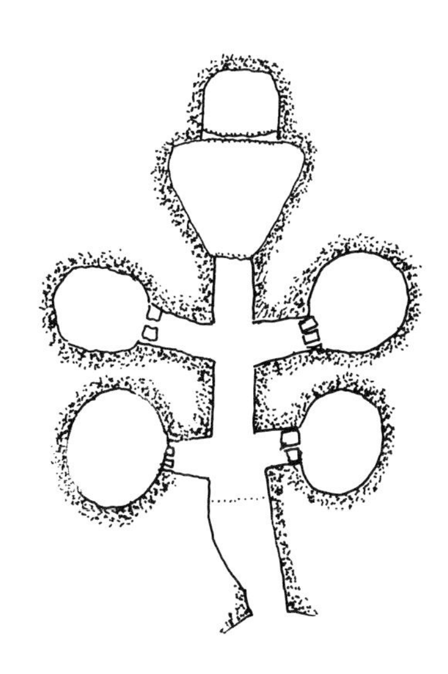

92. *The shapes of Pantalica graves (see p. 342) point to an ever-so-tenuous connection with those of Malta's sanctuaries.*

93. *Headless standing female statuette. National Museum, Malta.*

puzzle, for they had no "knowledge of the old." Their tools—mere chips of stone—would make the strongest workman faint. The shaping of their womb-tomb of Hal Saflieni, especially, was laborious beyond description. For lack of metal tools, they *clawed* their way into the rock with, of all things, antlers. If their feat does not exactly boggle the mind, it sends shudders down the spine. We shall probably never know what drove these people to undertake such constructions—surely not the plaudits of posterity. *We* certainly don't care for the approval of people several thousand years hence because we don't want to think what these monsters will be like.

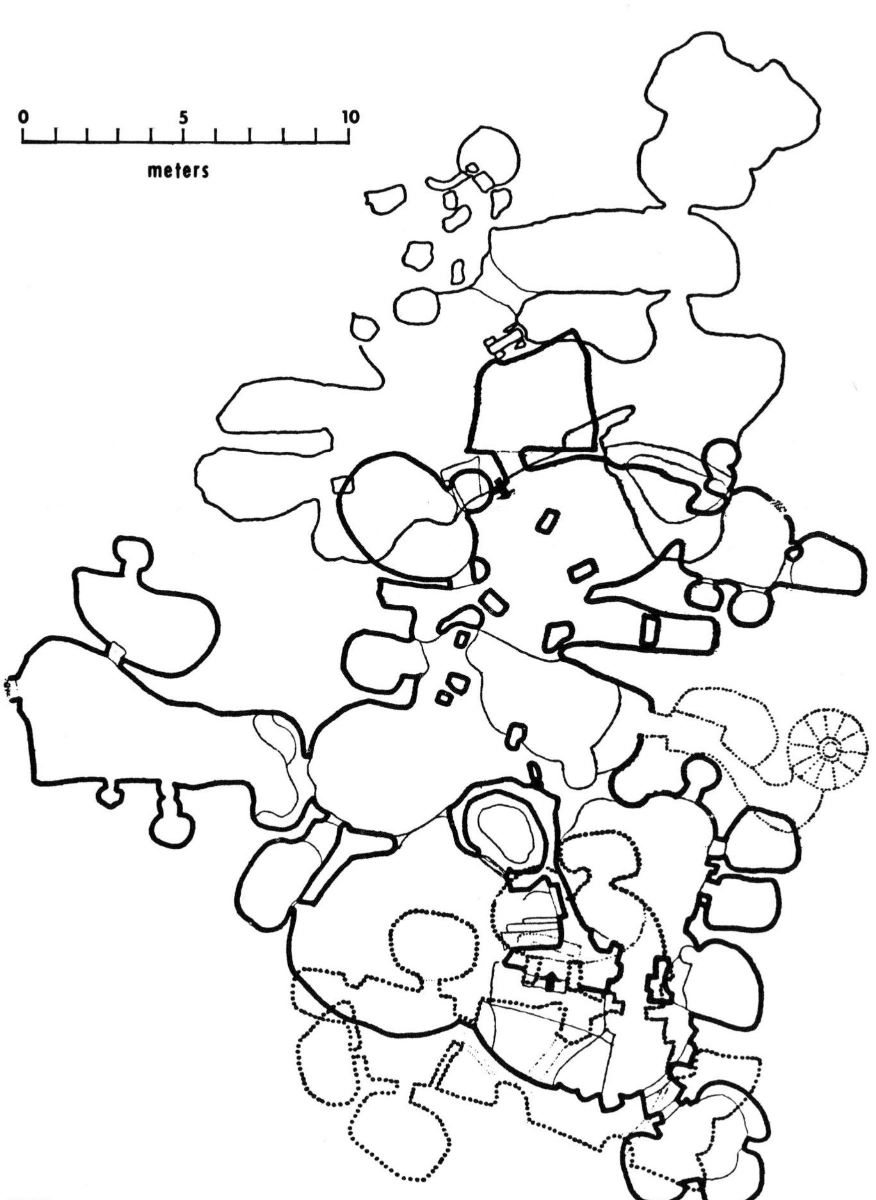

94. The Hypogeum's plan no more than hints at the complexity of its chambers and passages—labyrinthine spaces innocent of level floors and ceilings, perpendicular walls, and standardized steps. From Saggio sul Labirinto *by Michelangelo Cagiano de Azevedo. (Courtesy* Vita e Pensiero)

95. The sculptured wall surface of a near-circular hall in the Hypogeum imitates built-up megalithic architecture. Doors between the mock pillars lead to adjacent rooms. (Courtesy Malta Government Tourist Board)

Mobile architecture

To see the world without leaving home—and I don't mean on the flickering screen—is just one of the amenities that contemporary architecture does *not* provide. Our houses are irrevocably grounded. Sedentary man never gets a chance to sample such unconventional shelter as houses on wheels, sleigh huts, or floating villages that lent a high note of adventure to the domestic architecture of the past. In fact, he probably never heard of them. Only in a casual way does he opt for a house on wheels: Industrial society has affectionately invested the automobile with the status and some of the functions of a second home. So far, however, it lacks a hearth and an inhabitable floor, appurtenances essential to a reputable household.

Trailers are the nearest approximation to a free-wheeling domicile; yet, like most modern vehicles, they, too, are roadbound. Their range of mobility does not extend beyond highways and parking lots and perhaps for this reason they are rarely on the move. Whereas no trooper would dream of spending his days in the saddle of a horse tied to its crib, most trailerites never as much as contemplate going on the road. Being kept on that unbreakable leash twisted from tax liens and credit cards, the trailer serves them as a permanently moored home. The trailer park—a euphemism, applied as a rule to a barren camping ground—becomes the final resting place for what was originally intended as a means to escape sedentariness.

Mobile dwellings of generous proportions, suitable for cross-country travel, once were indispensable to nomads. Houses on wheels, for example, were a regular commodity of the eastern Scythians—the equestrians of the steppes—who perpetually moved back and forth between Danube and Don, and sometimes beyond. In Aeschylus' words, they "dwelled in wattled huts in the air upon their fair-wheeled wains." The huts were round or rectangular and divided into two or three rooms. Their walls were formed of basket work or of wands strung together with thongs. Layers of clay or felt made them proof against rain and snow. The small houses moved on four wheels, the large ones on six. They were pulled by several yokes of hornless oxen.

Herodotus, who visited Scythia on his way to Persia and is responsible for our habit of applying the name Scythians to all inhabitants of southern Russia, betrayed something like envy for their independence. "Having neither cities not forts," he wrote, "and carrying their dwellings with them wherever they go; accustomed moreover, one and all, to shoot from horseback, and living not by husbandry but by their cattle; their wagons the only houses they possess, how can they fail of being unconquerable?"[1] Savage in war, in peace they indulged in their cowboyish pursuit of capturing and taming wild horses. Life without a horse was as unthinkable to them as life without a car is to us. Even in afterlife they craved a horse. An important man might take to his grave a hundred horses, never minding the expense. An exemplary custom, it is worth imitating. For what could be more

96. *These Mongol tents consisted of several layers of greased felt stretched over lightweight wooden frames. The largest of them were mounted on wheels and pulled by twenty-two oxen. From* The Book of Ser Marco Polo *by Henry Yule.*

97. *This engraving of the curious mobile dwellings was copied from a drawing made on the spot by the Friar Rubruquis on his visit to Tartary in 1253. If, as has been conjectured, such wagon houses served as prototypes for Seljuk mausolea (see opposite page), this would be another instance of nonpedigreed architecture making the grade. From Rubruquis's* Voyage en Tartarie. (*Courtesy British Museum.*)

appropriate than to inter our great enthroned on a hecatomb of cars. Despite the mounted Scythians' rootlessness and seeming improvidence, their existence was anything but precarious. Being partial to horseflesh, they were able to carry their entire provisions on the hoof, so to speak. They milked the mares and made milk into cheese or fermented it into koumiss, an alcoholic beverage.

We may take comfort in learning that a strenous life, sustained by a high-protein diet, does not necessarily produce a healthy breed. No less a man than Hippocrates, the father of medicine, noted that the Scythians' way of life had serious drawbacks. (Like Herodotus, he had a first-hand knowledge of their habits.) He reproached Scythian kids for spending too much time in the family wagon. For they rarely walked, and neither did their fathers. A surfeit of transportation, Hippocrates maintained, sapped the men's strength and made their body fleshy and flabby; "the constant jolting of the horses," he noted, "unfits them for intercourse."[2]

It can't have been all that bad. During the centuries when the Mediterranean commonwealth slowly crumbled into ruins, wagon dwellers were much in evidence. Their names and ethnic identity changed but not their customs. In Asia, Hippocrates remarked, everything grows to greater beauty and size than elsewhere, and the wagon houses were no exception. Within a thousand years they had grown to be 30 feet wide, no mean size considering that the width of the average American townhouse is 5 feet less. "I took the trouble of measuring [the wagons] which were twenty feet from wheel to wheel, the sides of the house sticking out at least five feet on each side," wrote a traveler to Tartary.[3] The axle was as thick as a ship's mast, and the wheels of solid wood 12 feet high, a good deal taller than wheels used in modern carriages. None of the chroniclers tell us how the wagon houses, resting as they did on springless, overextended axles, performed on the roadless plains.

Despite their enthusiastic disregard of sedentariness, the Tartars by no means deprived themselves of domestic amenities. With one or two dozen wives on their hands—each of which had her own retinue—they lived off the fat of their women rather than of the land. "The court of one of these rich Tartars," wrote our informant, "resembles a market town where nevertheless there are few men. But the least of the wives would have twenty to thirty wagon houses for her attendants."[4]

98. The supposed models for tombs such as the Döner Kübet, built in 1276 at Kayseri, Turkey, were Tartar tents.

This polygamous arrangement was dictated by economical considerations. The men being busy with hunting and practicing archery, it was the women's lot to do the work. Their sphere of activity was all-embracing. They were in charge of the animals; they put the oxen and camels to the wagons and led them on the way. They made butter and dry milk. They fashioned furry garments and shoes from the skins they cured (they themselves wore trousers). And as with birds, among whom it usually is the duty of the female to make the nest, the women built the wagon houses.

Armies patterned themselves in many respects after the wanderers of the steppes. The Roman traveler Pietro della Valle, who accompanied the Persian Shah Abbas on a military campaign in 1618, was full of admiration for the resourcefulness of a soldiery constantly on the move. "They have invented a thousand ways to stay comfortable and to enjoy every convenience expected in a town," he wrote. "Thus they have portable baths which in camp they set up under their tents. And many times have I seen camels carrying big wooden installations for their baths that, I think, are used either as floors through which the bath water runs off without inconveniencing them, or serve to some such effect. Similarly, they have portable kitchens, and I don't mean pots and pans and kitchenware of the kind that just anybody would carry with him, but a kitchen range complete with tools, mounted on a camel so that one can do the cooking on the move."[5]

This early prototype of a *wagon-restaurant* has a truly spartan counterpart in southern Mexico. Nobody knows how far back originated the custom of cooking a meal on the run, but it seems to have always been a woman's privilege and duty. To put the time she spends walking home from work to good use, she carries on her head a brazier full of glowing coals plus a pot with its bubbling contents. Dinner is ready to be served as soon as she reaches her house. A prime example of economy, it illustrates folksy know-how at its best. City-bred people have restricted all headwork to the inner resources of their skull; the use of the head for bearing loads they have never known despite the example of that noble kind of architecture where caryatids and atlases are permanently upholding tons of pediments.

It takes but four people to balance a roof of modest size on their heads and to transport it from one place to another. Livingstone, traveling in the country of the Balonda, came to appreciate such movability. "When we had decided to remain for the night at any village," he wrote, "the inhabitants lent us the roofs of their huts, which in form resemble those of the Makoholo, or a Chinaman's hat, and can be taken off the walls at pleasure. They lifted them off, and brought them to the spot we had selected as our lodging, and, when my men had propped them up with stakes, they were then safely housed for the night."[6]

No wonder that to the natives a Western-style house seemed a paragon of grossness. Some tribesmen who had visited Livingstone's quarters at Kolobeng, in trying to describe them to their fellowmen, hit on a metaphor worth its weight in gold. To their mind it was not at all a house; "it is," they said, "a mountain with several caves in it."[7]

The portable shelter par excellence is the tent. It takes us back to just another of man's many attempts at a roof. Its starting point was a tree, except that the derivation is straight, without a detour over symbolism. Strange to say, among some peoples the word for *tree* and *house* are the same and, to a certain extent, so are the things the words stand for. For example, to pitch tents for winter quarters the ancient Argippaei stripped a live tree of its leaves and clothed the naked branches with felt—a feat of human ingenuity that takes its place with the Gordian knot and the egg of Columbus. They probably got the idea of the tree as umbrella from the *sedentary* Scythian

99. *The sectional view of a Dinka hut, about 40 feet in diameter, discloses an umbrella construction, the tree's branches supplying the spokes. The roof consists of layers of cut straw. Upper Nile. (After Schweinfurth)*

tribes of which Herodotus says that "they dwell each man under a tree, covering it in winter with a white felt cloth."[8]

They have their contemporary counterpart in the Altai Turks who remember this poetic genesis of the tent in a sacrificial ceremony that takes place in the clearing of a wood where they erect a special yurt from whose smoke hole projects the leafy top of a young beach. From these modest beginnings tents grew into palaces that defy modern man's imagination.

Nomads distinguish between tents according to size, shape, material, and use. There are words for flat-roofed and ridge-roofed tents; peaked ones like the tepee of the American Indians; umbrella-top tents and eight- or ten-sided ones; tents with canopies and porches that incorporate doors and windows; tents supported by a single pole; and others hung from three dozen masts and rigged with five hundred ropes. "In plan and structure," writes A. U. Pope, an authority on

100. The tents on an Assyrian bas-relief betray their affinity to prehistoric houses whose roof was supported by a tree in the center (see opposite page). (Courtesy British Museum)

Persian art, "a great tent might become so architectural that it would appear to be a castle."[9] He justly complains that "no European historian of art has ever seriously considered tents and pavilions as architecture."[10]

The common material for a tent's walls was, and still is, felt made from black goats' hair. Hides sewn together had the advantage of admitting a gleam of daylight. For more luxurious constructions, textiles, plain, patterned or embroidered, were used. In China a screen of

colored silk was put around the tents, like "the wall of a town."

Europeans got their first intimations of grand tent architecture from Marco Polo's travel accounts. The tent of the emperor of all Asia, Kublai Khan, was of such vastness that it put to shame Christendom's churches and palaces. Although we are not given its measurements, we are told that the tent in which the Khan gave audiences held ten thousand soldiers and still left ample space for persons of rank. Its imposing size was not diminished by the thou-

101. Military commanders pitching camp instinctively dispose their tents in the image of a town. Here, the town walls are symbolized by a circular tent structure. Detail of an engraving by C. N. Cochin, one of a series depicting the victories of the Chinese Emperor Ch'ien Lung. (Courtesy Prints Division, New York Public Library)

sands of tents that surrounded it. "A spectator," Polo wrote, "might conceive himself to be in the midst of a populous city."[11]

Yet bigness was merely a concomitant symptom of things Asian. Where the Khan left Barnum and Bailey far behind was in his tents' quality. If nonpedigreed architecture is generally on the frugal side, here it was of a mind-boggling extravaganza. Today's Croesuses would think twice before ruining themselves with similar expenditures. "The tents," noted Polo, "are covered on the outside with the skins of lions, streaked white, black and red, and so well joined together that neither wind nor rain can penetrate. Within they are lined with the skins of ermines and sables. . . ."[12] If the tent poles were of carved and gilt wood, and the tent ropes of silk, this was no doubt because more precious materials were not available.

Like the cave, the tent received divine recognition. In German, the

102. This picture of an airborne house flying the American flag in the pages of a nineteenth-century Chinese periodical may have been merely intended as a rapturous, if premature, tribute to Yankee ingenuity. (Courtesy Columbia East Asian Library)

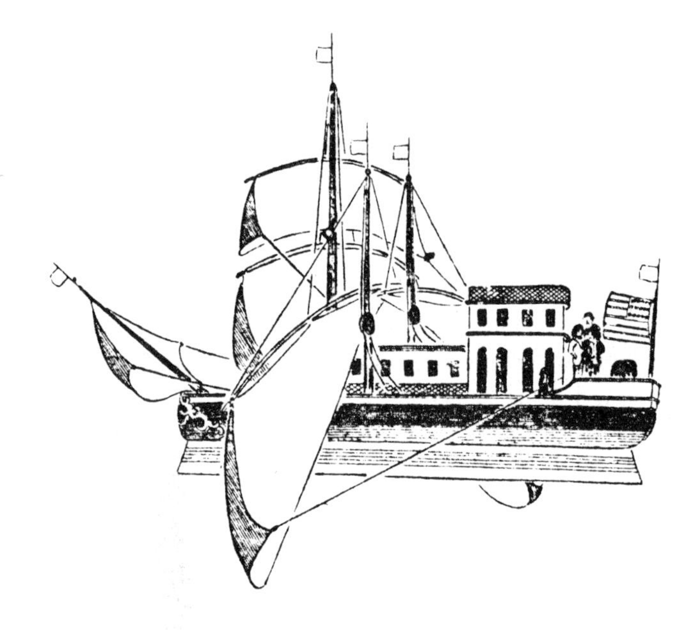

Himmelszelt, the heavenly tent, is no more than a figure of speech, and so is the English vault of Heaven (no allusion to Cosma's barrel vault). In our religious art, the myriads of Ascensions, Assumptions, and Last Judgments, Heaven is represented as a pall of stratocumulus, sometimes crowded to suffocation with angels and saints. However, it was the Muslims who brought a semblance of architectural order to it; they propounded a heavenly space concept that science took long in catching up with. In the Apocalypse of Muhammed, "the Prophet perceives, close to the throne of Allah, seventy thousand tents, each as large as the world, separated from each other by the distance of 70,000 years, in each of which are 50,000 angels who adore Allah."[13] Infidels whose powers of imagination fail in the face of such splendorous mega-architecture may prefer to take note of a more down-to-earth subject such as kites.

103. Before the invention of captive balloons, manned kites served as mobile observation towers. To lift this Japanese monster kite into the air required two hundred men.

104. In a fishing village of Japan's Inland Sea the bamboo frame of a kite is being assembled on the ground for the annual kite-flying festival.

Can kites be classified as architecture, if ever so fragile, without doing violence to linguistic usage? It probably depends on the kind one has in mind. Surely, *giant* kites have a claim to be considered architectural borderline cases. A 4-ton kite is not a toy, nor could flying it be called child's play. Yet if it is not a toy, neither is it a vehicle; one cannot travel by kite. By stretching a point, one might call it shelter, since occasionally it did accommodate a person. Aerial spotting was practiced long before the advent of flying machines, and among the earliest uses of kites was that of a lookout tower—a tower of variable height, without walls or stairs, a dancing, tossing belvedere on a leash.

In the Orient people have been flying kites since time immemorial; the Greeks only reinvented them. Long before Benjamin Franklin employed a kite in his blasphemous experiment (1752), the Chinese had assessed its practical application. However, like many other inventions, kites did not come into their own until their usefulness in battle had been proved. Kites capable of carrying a man literally unfolded new perspectives in strategy. Not only did the man aloft gain a bird's-eye view of the goings-on in an enemy camp, he was able to take potshots with impunity—a splendid example of soft technology.

When Marco Polo visited China, kites strong enough to lift a man were quite common. The Italians, ever avid for innovations, took a leaf from Polo's book and resolutely flew manned kites, made in Rome. However, with Mongolfier's invention of the free balloon (1782), the semistatic kites fell into oblivion, and it was only a century later that they recovered some of their prestige. Among the latter-day promoters of passenger kites stands out B. F. S. Baden-Powell—not the Father of Boy Scouts but a captain of the Scots Guards—who in 1894 successfully launched a man by means of a 36-foot-high kite, a variant from the parent stock. He even succeeded in raising one during a dead calm by adding half a dozen smaller kites to it and towing it along. Similar ascensions were conducted about that time in the United States.

In the Far East, where the fascination of kites never palled, annual flying contests are still held in a number of countries. But today's entries do not compare with those of the past. Japanese kites of old were truly castles in the air. Known as the Japanese are for their frugal tastes, their grandiose kites are all the more remarkable. The mere assembling of the building materials was a monumental chore, not undertaken lightly, and the expense involved put the building of big kites out of an individual's reach. The kite shown in figure 103 measured more than 50 feet across and required a construction site equal to that of several houses. Its oval skeleton was laid out, knotted together, and papered on the ground, and the finished work of art carried to a nearby beach where strong winds could be counted upon to blow on summer days. It took two hundred men to lift one into the air.

105. Part of the ground crew necessary for lifting one of the big kites into the air.

106. A squad of mobile towers constituted a maneuverable fortress, a great improvement over front-line elephants. The engraving depicts the siege of Jerusalem by the Seleucid King Antiochus VII Sidetes in 134 B.C.

107. Right: Gravure of an assault tower from the Gujin tushu jincheng, *an encyclopedia compiled at the order of the Chinese emperor Kangxi. (Courtesy Columbia East Asian Library)*

108. Opposite page: The Trojan Horse, a zoomorphic variant of the ambulant towers on wheels used in laying siege to a town. German Archaeological Institute, Athens.

The winds still blow as fiercely as in 1281 when they sank Kublai Khan's armada that carried the Mongolian forces to the invasion of Japan. However, the spirit that produced the monster kites has given way to a passion for passive entertainment. Moreover, while bamboo of uncommon dimensions and rope as thick as hawsers can still be found, nobody makes paper as strong as sailcloth anymore. The last giant kites were flown in 1914, whereupon they joined the extinct subspecies of architecture such as rolling towers and swimming fortresses.

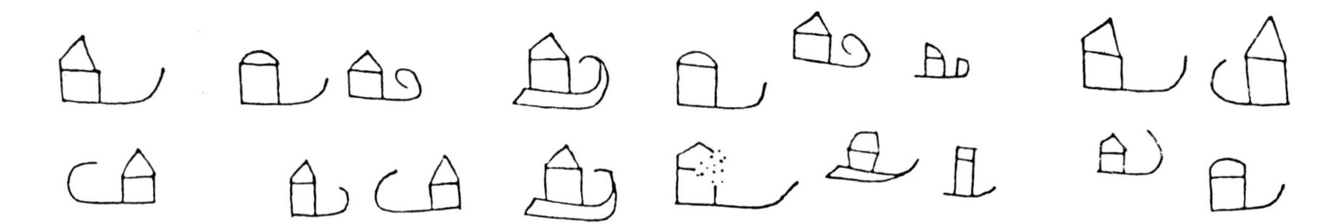

109. Pictographs representing sleigh houses, culled from Babylonian inscriptions circa 3500 B.C. *From* Archaische Texte aus Uruk *by Adam Falkenstein.*

A variant of mobile dwellings, uncharted on the maps of architecture, are sleigh houses. A *sleigh* is defined as a vehicle on runners for traveling over snow and ice. It was left to unsophisticated shepherds to include in this definition sleighs that travel over mud and meadows as well. Their runners are wider and stronger than those of ordinary sleighs; in fact, they are sturdy enough to serve as foundations for a family-size hut. Such sleigh or sled houses once were common enough in the Balkans, from Hungary all the way to Bosnia, Bulgaria, southern Serbia, and Greece.

To Americans the Balkans may seem more remote than the moon, a sled house no more real than a pumpkin coach. They may be surprised to learn that sled houses once were part of the American scene. While such dwellings may have been of no particular interest to the native, they did catch the eye of the observant foreigner. In 1795, a German physician on a visit to the New World, walking through Philadelphia's countryside, came upon the hut of a charcoal burner which stood "on a sort of sleigh, in order to be moved from place to place according to the occasion."[14]

Today, sleigh houses are enshrined in museums, but a few can still be seen on the plain around Sofia, the temporary habitat of the Vlachs, or Wallachians. Nomadic shepherds by choice, they do not cultivate the land but wander with their flocks in search of fresh pastures, and such is the climate that in summer they find grazing grounds in the mountains and in winter never run out of them on the plains. Their pace may be slow, but then, neither sheep nor shepherds are in a hurry. Like the ancient wagon houses, the sleigh houses are pulled by oxen. Three or four pairs, hitched to the runners, ride over terrain that otherwise can be negotiated only on foot, on horseback, or by jeep. (Not until 1908, when an Englishman, Hornsby, fitted a high-powered car with tracks, did a modern vehicle demonstrate a capacity for roadless travel.) Compared to a jeep, however, a sleigh house is spacious. There is enough room for storing milk and cheese and all the paraphernalia of the trade. In a way, these sleigh houses are the mobile counterpart of Swiss *Sennhütten*, or alpine dairies.

The largest of them, called "palaces" by the peasants, measure as much as 10 by 12 feet. A thatched roof, walls of wood or woven reeds, a low door, and occasionally a window is all there is to it. However, the structure is solid enough to permit cooking in the center of the room. For lack of a chimney, the smoke escapes through the door. Around the rude hearth is sleeping space for four people. Unlike Basque shepherds, the Wallachians are no loners; they like to be with their families.

110. Above: Man has always taken pleasure in speed and mobility. Skiing, introduced in America barely a century ago, was practiced by prehistoric society in sub-Boreal times. (After Gurina)

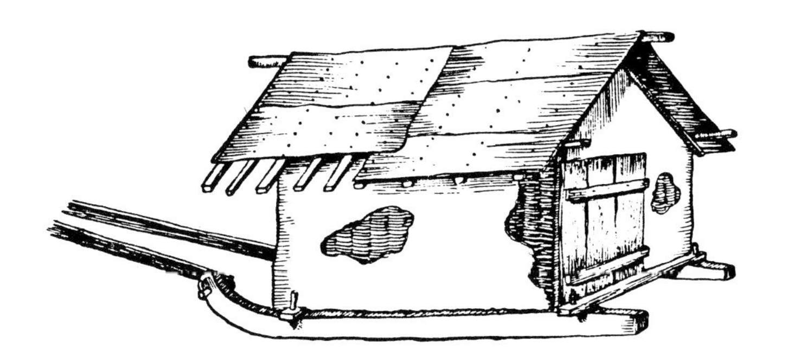

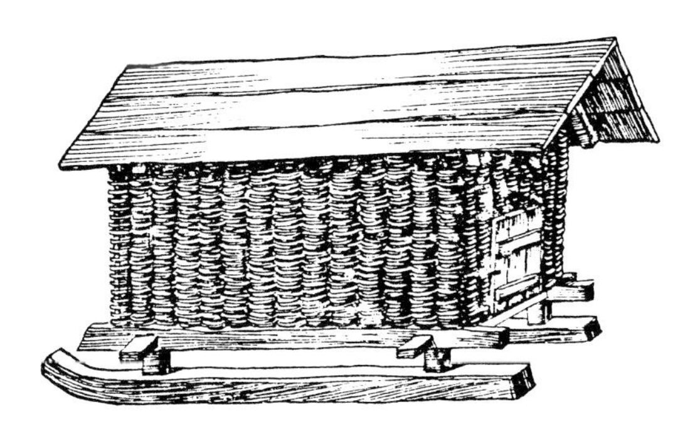

111. Sleigh huts of Bulgarian nomadic shepherds house entire families. The patches on the walls and roofs in the drawings are not picturesque touches but are meant to show the huts' construction. From "Die Bulgarischen Hirtenhütten" by Khristo Wakarelski, in Acta Ethnographica.

112. Wooden wheel of a Portuguese mill near Leira. The entire mill rather than its sails turn with the wind. From Arquitectura popular em Portugal. *(Courtesy Sindicato Nacional dos Arquitectos, Lisbon)*

113. Revolving windmill with wooden sails. Herăstrău, Rumania (see pp. 277 ff.).

Togetherness on a more impressive scale prevailed in those least-known—indeed, unheard-of—communities, floating villages. Whereas in our country a houseboat counts among life's luxuries, in the Far East some of the most impecunious people live permanently on the water. They take to every sort of craft, and congregate in protected bays, on lakes and navigable rivers. Many are born and brought up on ships and floats and hardly ever set foot on land. Yet although agglomerations of hundreds of houseboats are nothing out of the ordinary in some of Asia's large towns, amphibious villages disappeared long ago. The following eyewitness account is fully three centuries old:

"We saw," wrote Johan Nieuhof, an official of the Dutch East India Company, "upon the Yellow River, which is continually plowed with all manner of great and small vessels, several floating islands, which were so artificially contrived, that the best artists in Europe would but coldly be able to make the like of the same stuff; a common reed which the Portuguese call bamboo, twisted so close together that no moisture can penetrate."[15] The kind of bamboo used for building unsinkable rafts is of giant size—the largest species attain a height of 120 feet—with a buoyancy surpassed only by balsa and cork. A carrier with a cargo of 7 tons draws as little as 3 inches of water and never more than 6.[16] "Upon these reeds," Nieuhof continues, "the Chinese set up huts, like little houses of boards, and other light materials, in which they live with their wives and children, as if they had their dwellings upon the firm land: Some of these floating islands are able to contain at least two hundred families, and those that live in them subsist for the most part by commerce and traffic

114. "Flowing village" from Francisci's Lustgarten *(1668) presents but a modest Chinese hamlet afloat on the Yellow River. Similar itinerant communities of as many as two hundred families were nothing out of the ordinary in former times.*

upon the river; they are hurried down the stream, or towed up by toilsome bargemen."[17]

Nieuhof, who spent a year in China negotiating trade relations, had an eye for uncommon local aspects and seems to have been a reliable reporter. His observation that "they keep and feed aboard their island all manner of tame cattle, especially hogs,"[18] is supplemented by an Austin friar, Gonzáles de Mendoza, who noted that they grow flowers, vegetables, and orange trees on their floats. Where on earth does one come across anything like these nautical villagers, devoted simultaneously to stock breeding, gardening, fishing, and the trading of goods!

Two hundred families, each forming a populous clan, occupied no doubt a lot of space. Besides, each float may have boasted a temple of sorts that doubled as meeting place and general focus of life. Certainly, there were any number of storehouses, stables, and pigsties, unless the pigs were allowed to freely roam the amphibious village square. One would like to know more about the life of these river people who navigated the mighty Huang-Ho, then one of China's main highways. Did the floating villages ever come to grief? Although impervious to the dangers of earthquakes and floods, might they not have occasionally sprung a leak or been dashed to pieces on a hidden rock? Alas, no detailed reports on the itinerant communities are extant.

The unwholesome blend of a hazy past and a muffled present defeats all attempts at appraising some of architecture's more unusual aspects. If Nieuhof was no Marco Polo, neither was he a Münchhausen. His floating villages may be too much for our imagination, yet they are no fantastications. The Chinese, masters of the grand design, were capable of much bolder feats. The waterborne communities of the Huang-Ho had their counterpart, only far more imposing, in China's waterborne castles. The long-reigning Emperor Wu Ti (140–86 B.C.), making ready for the warpath, had a wooden fortress built on a square float, its sides measuring 600 feet, or the width of twenty-four Manhattan brownstone houses, and its walls strengthened by high towers with four gates for sallies. This buoyant castle was garrisoned by more than two thousand men, not counting the horses that romped the ramparts.[19] Withal, this was no dreadnought. As a work of art and artfulness, it was more akin to the zoomorphic fortress on wheels, the Trojan Horse.

Again, no pictures are extant. Only from illustrations of Occidental, small-scale swimming forts—mere knick-knack compared to Wu Ti's naval citadel—can we form a mental picture of the Chinese hybrids. Such scarcity of representations is deplorable, the more so because floating architecture has an honorable, quasi-sacred history.

Its oldest example is Noah's Ark. That grandiose project, next in line after Enochtown, involved but a minimum of human initiative and know-how since it was commissioned and built to the Lord's specifications. Contrary to general belief, foisted upon us by cinematic heresies, the Ark was no ship. It had neither stem nor stern, nor was it propelled by oars or sails. Above all, it had no keel. Hence it was never properly launched, though Noah might have enjoyed bouncing a goatskin of his best wine against one of its corners. The Ark's function simply was to stay afloat. What it amounted to was a "swimming house,"[20] a rectangular box of gigantic size. Ships proper had yet to be invented.

To be sure, the Babylonians knew how to build primitive rafts of the kind still seen on the Tigris in our days: floats, buoyed up by inflated animal skins or bundles of reeds, or osiers held together by cords and waterproofed with bitumen and pitch. Wood, always scarce in the biblical lands, did not find much application as building material. Besides, wooden houses could have ridden out the Great Waters as easily as the Ark itself, a fact corroborated a hundredfold by every Mid-western flood. Obviously, the existence of any seaworthy craft would have defeated the very purpose of the Flood.

As every child knows, the Deluge did not take people by surprise. They had been forewarned of its coming one hundred and twenty years ahead of time by what is no doubt the most anticipatory weather forecast on record. Yet, just as today we are unconcerned about our extermination through modern technology, Noah's fellowmen chose to worry about lesser perils.

The question of what the Ark looked like has occupied the minds of theologians and laymen alike. Luther—who called it a *Kasten*, a chest—thought it had a flat roof, an opinion shared by the German artists of his time who used to depict the Ark as a rectangular box. The Hebrew word for the Ark, *tēbah*—from the Egyptian *teb*(*t*)—likewise means box, and in the New Testament it sometimes is referred to as a "coffer." One has to keep in mind that in antiquity chests had many uses, not the least uncommon of which was to set undesirable creatures adrift in them. The adventure of little Moses bobbing on the Nile in a wooden box is sufficiently known. Danae's son Perseus had a similar experience. Both came to a happy end, but sometimes the little passenger drowned or died of exposure.

Not wanting to entrust the Ark's planning to Noah—a tent dweller who probably had never seen a wooden structure—the Lord furnished him precise instructions. According to Genesis, the venerable kennel was 300 cubits long, 50 cubits wide, and 30 cubits high, the height to be divided into three floors. Conservatively translated, these dimensions run to about 525, 87, and 52 feet respectively. However, the

115. Opposite page: The floating fort, complete with bastions, tower, and turret, is a small edition of the waterborne castles of ancient China. From Della Fortificazione delle Città *(1563) by Girolamo Maggi.*

de Timmering der Arke onder't bestuur van de goddelyke Bouwmeester Noach

116. *Wonder-working goodspeed rather than technology may account for the successful construction and floating of this alarmingly massive Ark. To shore up his contemporaries' belief in the extravagant enterprise, the Dutch artist invested Noah with the pecuniary means and sovereign powers of a pharaoh. From* Galerie du Monde *by P. van der Aa.* (Courtesy Prints Division, New York Public Library)

ancient writers were anything but unanimous on this point; Berosus, a Chaldean priest and an authority on the Flood, believed that the Ark measured 3,000 feet, or three times the length of Cunard's late *Queen Elizabeth.* And the second-century Origen, the most distinguished theologian of the early Church, arrived at the length of 25 miles. A man of uncommon faith and vision, he questioned and promptly dismissed the measly figures given by the Bible. "Why," he asked, "should we not rather admire a structure which resembled an extensive city, if its measurements be taken to mean what they are capable of meaning, so that it was nine myriads of cubits long in the base, and two thousand five hundred in breadth?"[21]

Whatever the actual size of the Ark, its concept inspires awe. Although the narrator of Genesis is chary of furnishing details, the great body of theological writings more than makes up for what the

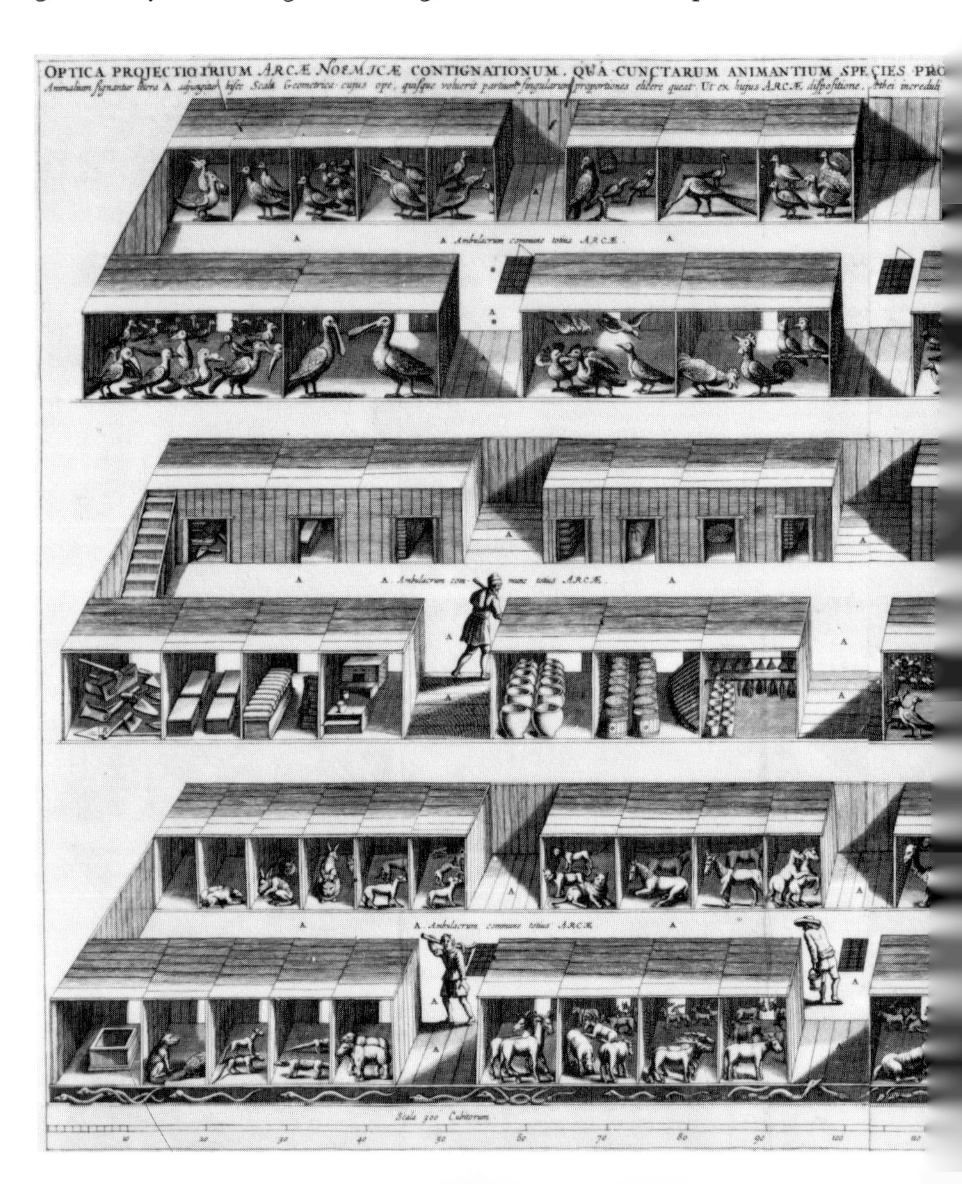

117. In his book on the Ark, Kircher shows the most spectacular frame house of all times with its full complement. From Arka Noe *(1675) by Athanasius Kircher.*

Bible withholds. Thanks to such erudite scholars as Father Athanasius Kircher—he called the Ark "one of the greatest edifices made by man"—we are well informed about it. According to Kircher, who devoted a weighty and generously illustrated volume to the description of the floating zoo, Noah's quarters were anything but cramped. They consisted of a bedroom, dining room, kitchen, and pantry, or the equivalent of a middle-income family's apartment. Next to it, and well within earshot, was an aviary filled with songbirds, probably the earliest instance of piped-in music. There were separate rooms for each of his sons and their wives and, equally important, separate cubicles for each pair of animals. Yet even with space fully utilized to the beams, the bestiary was necessarily incomplete. Divinely conceived as was the Ark, it could not accommodate colossal creatures. Were the centaurs, satyrs, fauns admitted? facetiously asks a modern

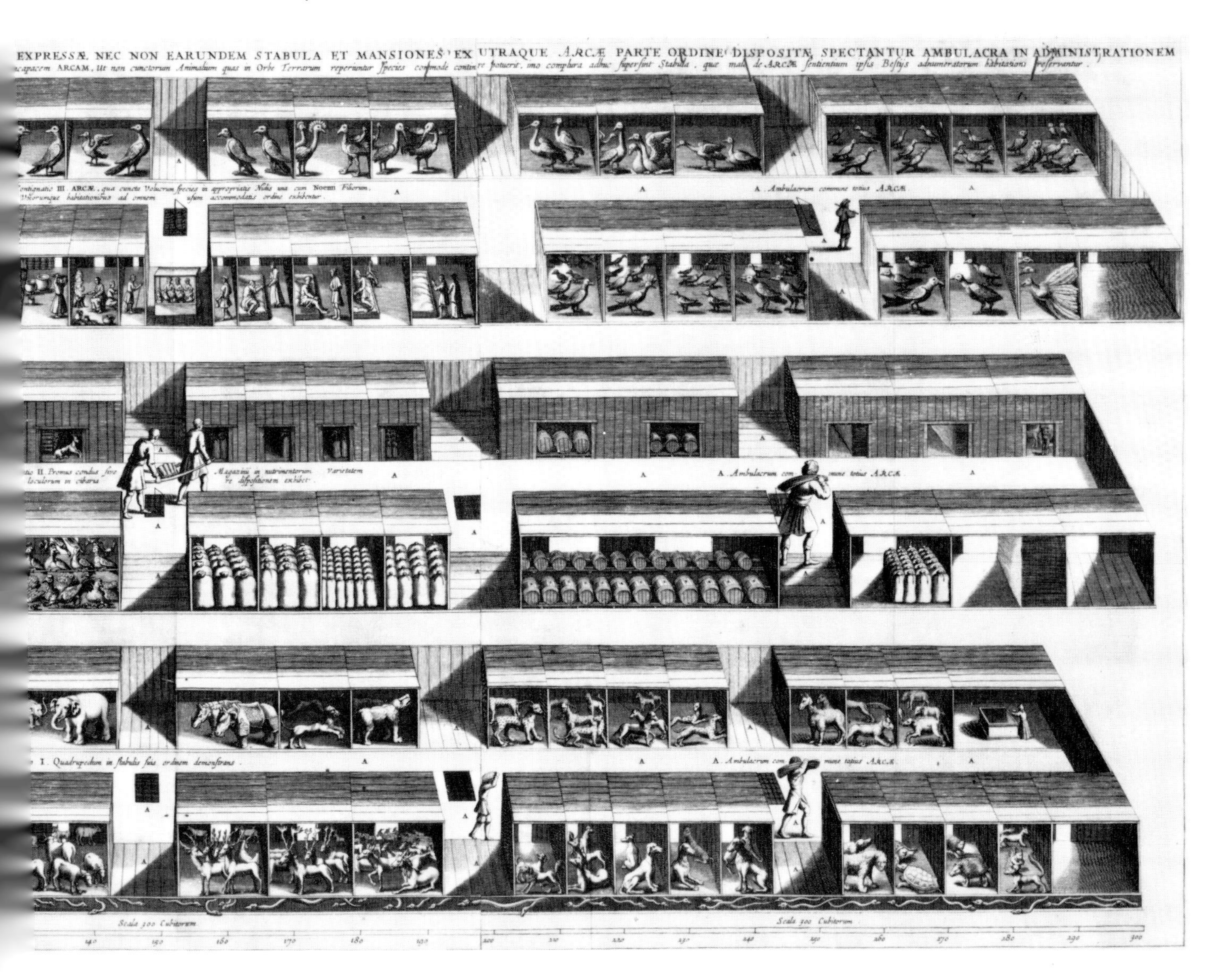

scholar.[22] Although on this point Kircher keeps silent, his answer might have been affirmative, for he kept the skeleton of a siren in his own museum. Since the stenches, human and bestial, cannot have but put a damper on the enjoyment of the cruise, Kircher thoughtfully added for Noah's comfort a rudimentary air-conditioning system, "special air vents that kept the atmosphere quite fresh."

Rabbinical literature fills in another interesting detail: An entire year spent in utter darkness—"the stars and planets did not fulfill their function during the Deluge"—might have driven the eight human passengers raving mad and endangered the beasts as well, had Noah not hit upon an ingenious and novel type of illumination, precious stones that shone "bright as the sun at noonday," a variant of the luminous wreath, or crown, which Ariadne gave to Theseus for shedding light upon the labyrinth. Alas, we do not know what kind of stones were used for those first Floodlights, since even diamonds depend on the outward boost of light to kindle their inner fire.

118, 119. Unsuited for navigation, Noah's Swimming House drifted 25 feet above the tip of the Caucasus—the high-water mark of the Flood—and, after skirting Athos, Olympus, and Taurus, landed (opposite page) on the top of Mount Ararat. (In the actual, Mesopotamian, inundation, which inspired the biblical myth, the waters rose to no more than 16 feet.) From Kircher's Arka Noe.

Kircher conjectured that Noah did not lend a hand in shaping the Ark but merely supervised its construction. Still, it seems unlikely that he enlisted his neighbors' help, which would have smacked of the practice of certain executioners who put the condemned to digging their own graves. (In the original story of the Flood from which the biblical version was lifted, Utanapishtim, the ur-Noah, had the decency to invite "all the craftsmen" for the ride.) Whoever supplied the labor, the Ark's workmanship matched the grand design. The roof sprang no leak, the caulking, specially mentioned in the Gospels, easily withstood a rainfall that exceeded 200 inches per hour, 400 feet per day. Oddly enough, the Bible says nothing about the difficulties of the Ark's landing—stranding might be a more fitting word—which could not have been achieved without the assistance of a major miracle. Mount Ararat, the Ark's terminal, is all lava and Archaean schists, in other words, as jagged as stone formations come, and the ponderous structure had to perform an absolutely even landing in

120. Generations ago, Parisians built floating bathing establishments on the Seine while New York's citizens, ever-anxious to cleanse the inner man, launched an authentic church upon the murky waters. 1844. (Courtesy Prints Division, New York Public Library)

order to avoid cracking up, spilling and maiming its precious contents.

The whereabouts of the Ark has at all times excited the curiosity of people with a lively imagination. Ancient and less ancient rumors would indicate that the Ark rests safely on the mountain. Indeed, its location was supposed to be known centuries before Christ. In his *Lustgarten* (1668), Erasmus Francisci tells that the Ark was visible in clear weather, but somehow people were not eager to inspect it at close quarters. Some did scrape off bits of its caulking for use as antidote to poison or infection, but the ascension of the mountain proved far too arduous a task to be undertaken lightly even by the most pious. To keep the Ark's memory alive, fragments of its hull were displayed and venerated in churches and convents. At one time a chunk of its wood was owned by the Echmiadzin monastery, the seat of the primate of the Armenian Church, while another piece found its way into the Lateran, only to vanish without a trace. What adds a special nimbus to these relics is the fact that they had not been picked up by shepherds turned souvenir hunters but had been delivered, as it were, by air mail, that is, bestowed upon some deserving holy men by angels.

Of all the structures erected by human labor, Noah's house upon the waters remains in a class all of its own. The only piece of architecture ever built according to the Lord's specifications, the only one instrumental in postponing mankind's extinction, it is also the one which least affected architecture in general. Its few revivals can be put down to false alarms and commercial enterprise rather than to inspired capers. In 1542, a Frenchman of Toulouse built himself an ark, an event that failed to unlock the sluices of Heaven. In 1604, one Peter Jensen, a Dutchman of Hoorn on the Ijsselmeer, made himself the laughingstock of his fellow citizens by following the biblical blueprint. He got the better of his scoffers by demonstrating the exceptionally large tonnage of his swimming house. Its angularity allowed for one-third more cargo than a ship of the same dimensions, an advantage that was not lost on the thrifty Hollanders. What spoke against it and condemned it to oblivion was its unsuitability as a battleship.

The last would-be Noah heard of (in 1931) was a citizen of Olympia in the state of Washington. Anticipating the destruction of the Pacific Coast cities by a modern Flood, he put his faith and money in an Ark, made in U.S.A.

121. Tomb of a sixteenth-century Muslim ruler of the Songhai Empire, West Africa, the present Mali. A stairway leads to the top of the truncated pyramid, whence one enjoys a view of the countryside.

Storehouses, sepulchral and cereal

"Ulysses in Hecuba cared not how meanly he lived, so he might finde a noble Tomb after death. Great Persons affected great Monuments."[1] So Sir Thomas Browne. This irrational preoccupation with a posthumous life style persisted through the ages among the rich and the poor, the exalted and the wretched alike. Thumbing through a book of architectural history, one is struck by the prevalence of sepulchral monuments. In countries like Lycia, Phrygia, or Etruria, tombs are often the only vestiges of the past, because the houses of the dead have always been sturdier than those of the living. In ancient Egypt life was held cheap; what people valued was the afterlife. The houses of the living they called lodgings to emphasize their temporary usefulness, while they looked upon the grave as their permanent home. The Romans shared this attitude and referred to the grave as *domus aeterna.* Similar sentiments prevailed among such remote peoples as the Incas. "It truly amazes me," observed Cieza de León, the Herodotus of the New World, "to think how little store the living set by having large, fine houses, and the care with which they adorned the graves where they were to be buried, as though this constituted their entire happiness."[2] Thus, the tomb represented all that architecture was about while mundane shelter remained on the whole a sketchy affair, as provisional as life itself.

122. Each of the pyramid's sides is 30 feet long. The sticks serve as permanent scaffolding. (See also fig. 222 on p. 262.)

Man's hankering for permanent moorings on this treacherous planet has not subsided in the course of time. In Mediterranean countries, for instance, a man's wish to enshrine his remains may be

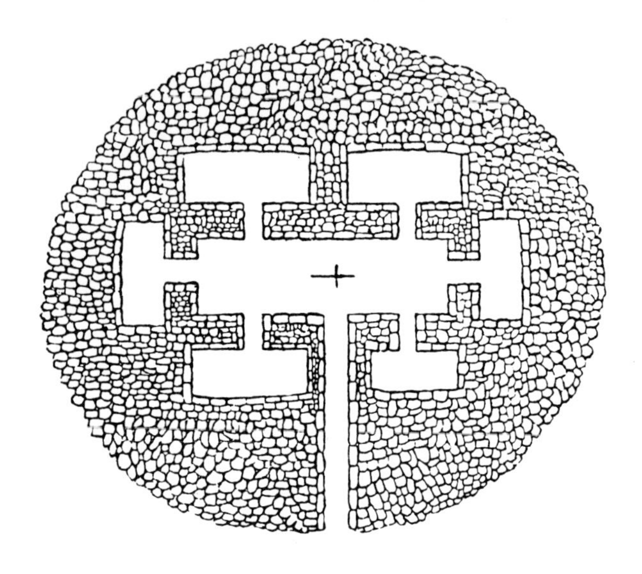

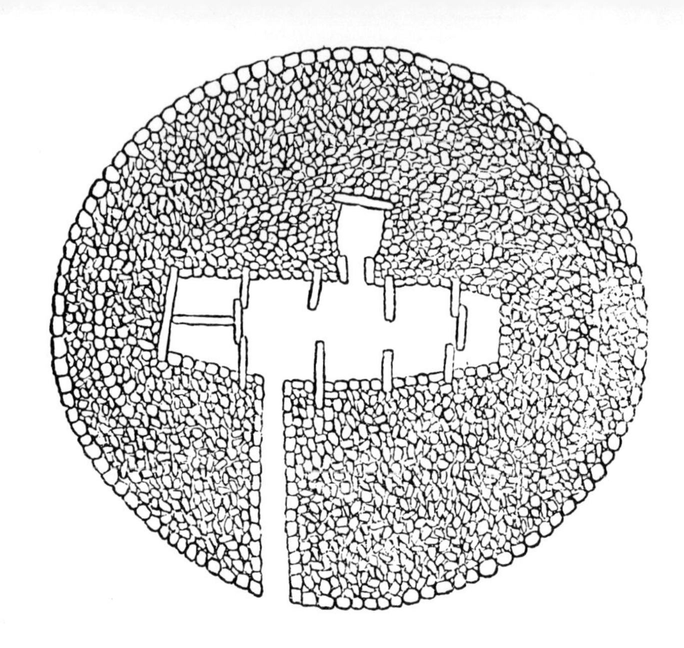

123, 124, 125. The three Stone Age graves in the Orkney Islands (above—Quanterness, right—Unstan, opposite page—Papa Westray) have come a long way from the time when a hole in the ground was thought fit to receive a corpse; they are truly houses for the dead. The largest of them is 100 feet long. Low corridors lead to a central chamber which is either subdivided by stone slabs or connected with lateral rooms. All spaces are covered with pseudovaults. From Orient und Europa *by Oskar Montelius.*

stronger than the desire for worldly possessions. Despite his zest for life, it is the prospective demise that occupies his most private thoughts. To this day, a well-to-do Neapolitan shopkeeper will forgo the pleasure of owning a house and rather spend his earnings on an impressive family vault. Neither a numerous progeny nor his fellow-men's love will give him as much inward satisfaction as the thought of a tomb that confers upon him, if not immortality, at least some measure of transcendental prosperity. He thus reenacts, on a scale commensurate to his means, the architectural fantasies of megalomaniacal rulers who aspired to perpetual rest in strongholds weighed down by artificial mountains, of which Egypt's Pyramids are a classical example.

Early man was exacting when it came to food and clothes only. The side of a bear was fine for a meal, and he probably did not feel effeminate in a long fur coat of sorts. His idea of protection may have been just a wrap—a portable rather than a static shelter. And there the matter rested for an unconscionably long time. To build an artificial shelter was as alien to his mind as eating artificial food. At all events, Stone Age man's first attempts at building were not motivated by self-indulgence. What prompted him to dig deep into the earth and pile stone upon stone was the wish to provide a home for the dead. He was very good at this, and would no doubt have been scandalized by the meanness of our burial customs. To commit the defunct to a narrow wooden box, and mark the site of interment with a cross or a niggardly stone slab might have struck him as the nadir of *savoir-mourir.*

Graves and tombs tell us more about mankind's needs than huts and houses. Moreover, while today's graves contain but a few bones,

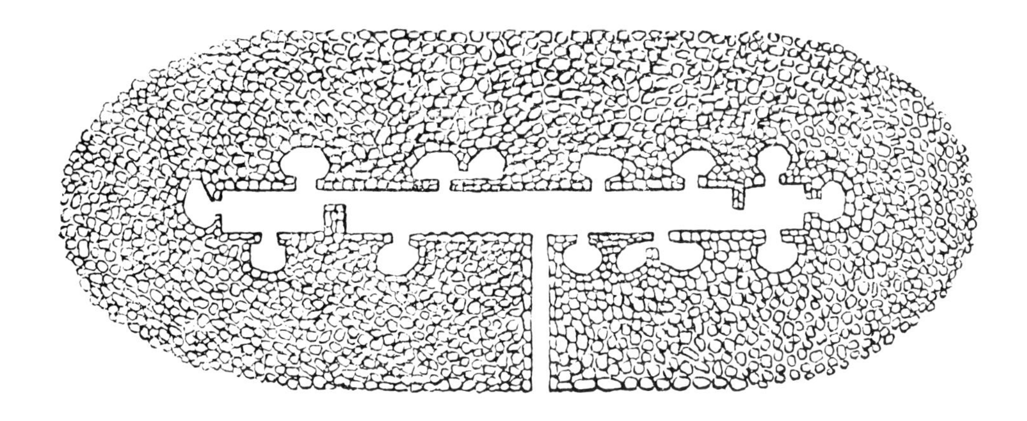

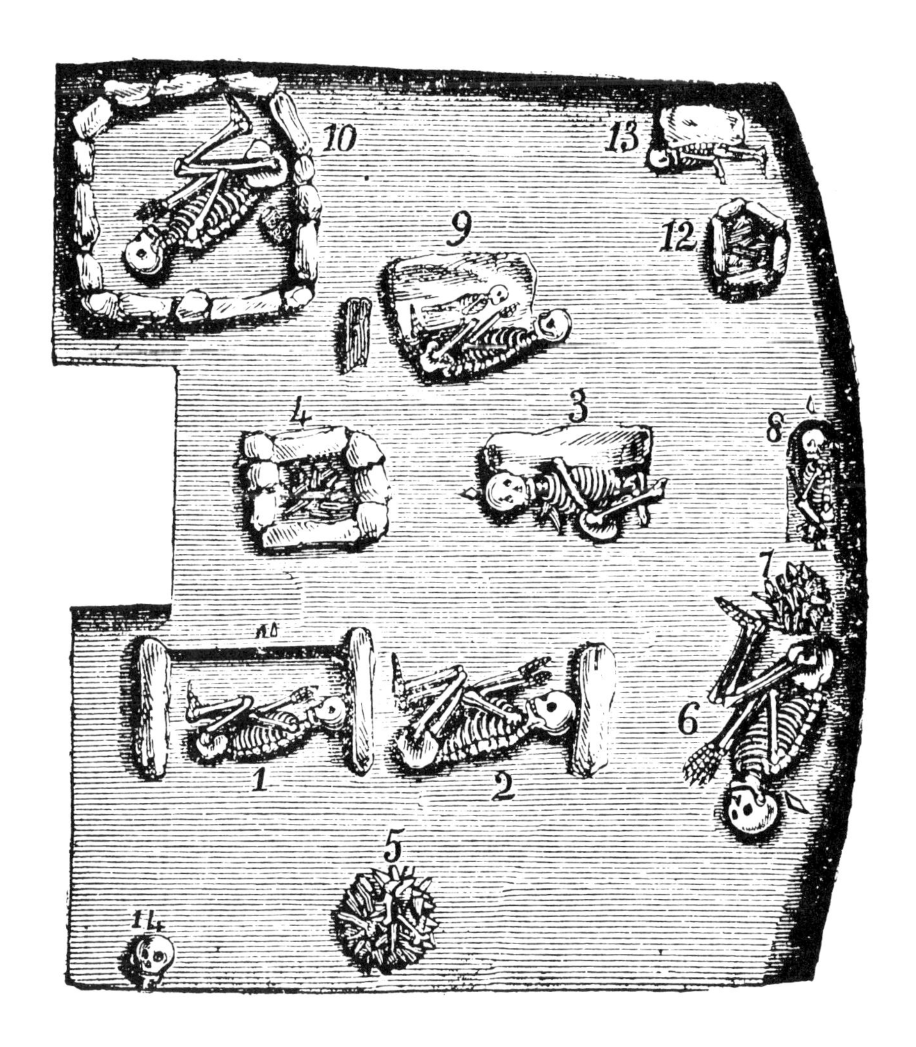

126. The contents of this barrow illustrate Stone Age sepulchral etiquette: The dead were put to rest in a contracted position and a stone on edge placed at their head or feet, or both. The bones of a young hog in a cist (at 4) add an unexpected note of geniality. From Gravemounds and Their Contents *by Llewellyn Jewitt.*

those of the past amounted to veritable treasure troves, glorified versions of the houses of the living that revealed the comforts of life in amazing detail. In many respects, tombs are the inventories of a civilization. Had some races left no tombs, we probably would not know that they existed.

Not that the dead always rejoice in their proverbially deep sleep. Ever so often they are subject to the same vexations that go with living in a flat, or cottage, such as high cost of the premises' upkeep; competition with the Joneses' next headstone; dread of the burial grounds running to seed; and so forth. Equally precarious is the existence of those poor souls reduced to inhabiting columbaria, the tenements of the neediest. They are left in peace only as long as the rent is paid. Failing this, their bones are evicted and their pigeonhole let out again.

There are as many ways of doing away with a corpse as there are of putting a person to death. All defy rationalization and not a few lie outside the sphere of our morality. Some people save themselves time and expense by leaving the undertaker's business to Nature, as when religious observances demand that a corpse be offered to vultures or holy sharks, and the leavings, if any, be placed in an ossuary. Such

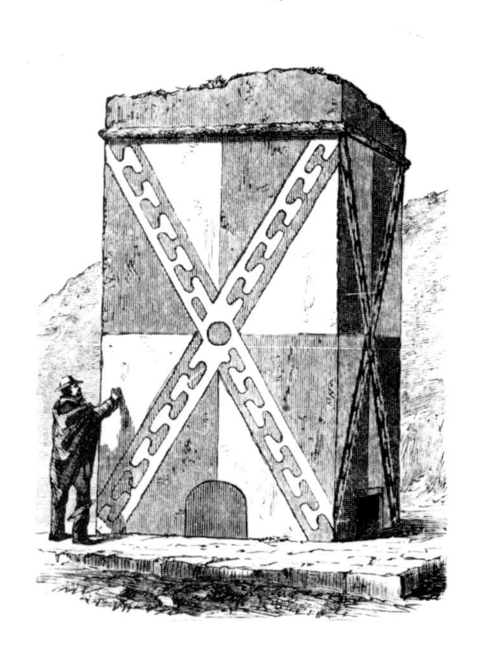

127. *Peruvian burial tower, 16 feet high, of solid stone, its surface smoothed and covered with fine clay, painted white and red. From* Peru *by George E. Squier.*

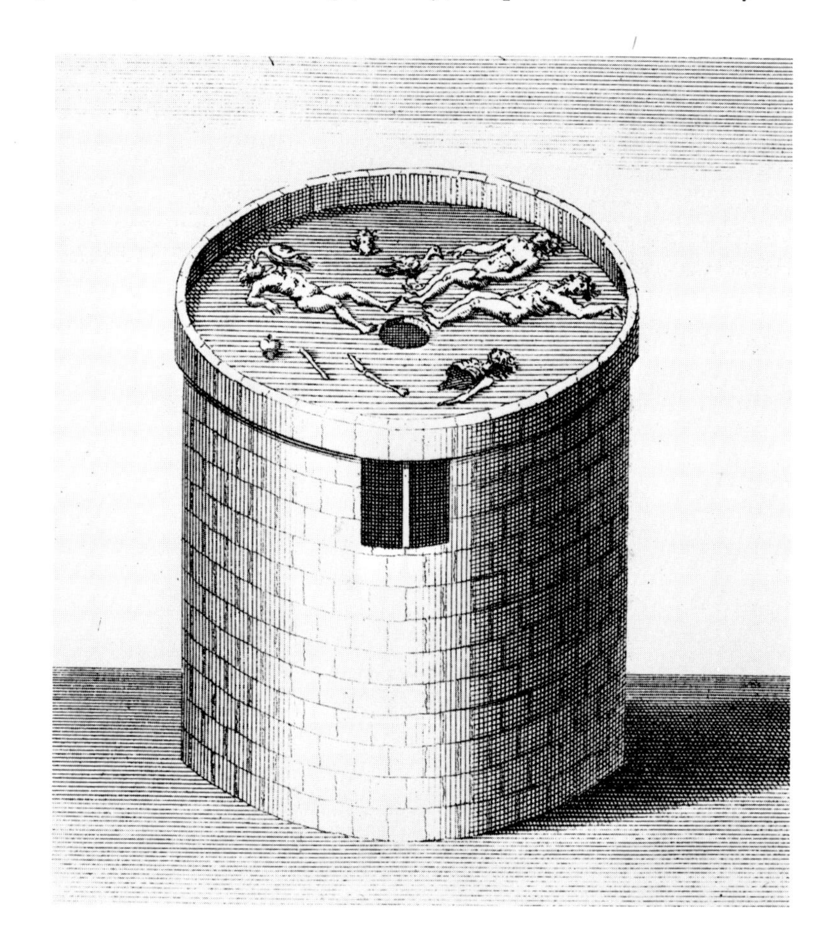

128. *Parsee funeral tower. From* Cérémonies et Coutumes religieuses *(1789). (Courtesy Lilly Library, Indiana University)*

customs have the added advantage of reducing all further care to a minimum. An equally economical solution is cannibalism. Tabooed in Western civilizations, it has however been sanctioned on more than one occasion. Another method obviating burial, popular with seafaring tribes, consists of sending off the dead in a canoe for an unspecified destination. Many consider this watery exit poetic, perhaps because it calls to mind Greek mythology where the souls of the deceased are ferried over Acheron and Styx. (The canoe is omitted in the case of slaves, foreigners, and people of no account.) Cremation is one more half-measure which does not solve the problem of what to do with the residue. Pleasantly associated with the funeral pyre, cremation is much in favor with the fastidious, if only because the

129. The dreary paraphernalia of crosses, tombstones, bad statuary, and funereal flora that make our cemeteries such forbidding places are absent from this hilltop kasbah of the dead in southern Spain. Loggie, courtyards, majestic trees, and a view of mountains and the sea add up to an agreeable retreat from life.

130. Loath to waste precious space for graveyards, the Spaniards bury their dead in high-rises reminiscent of African granaries (opposite page).

131. Storehouses at Medenine, Tunisia.

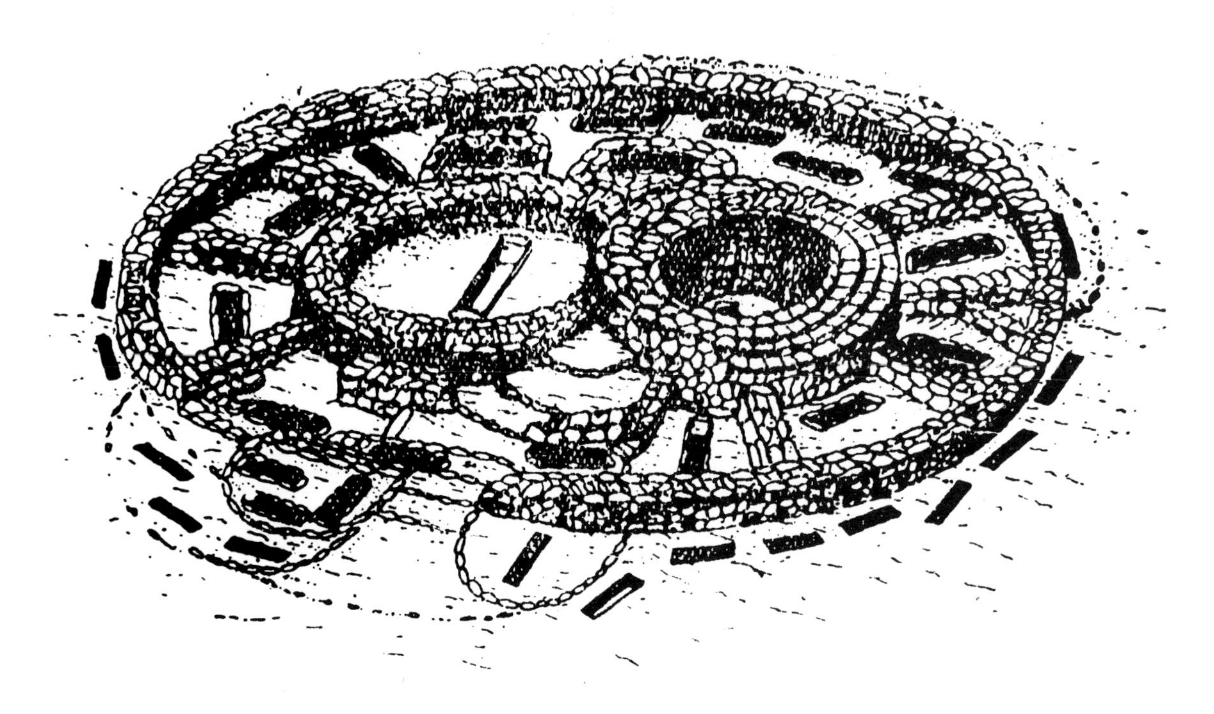

132. A tumulus built by the Guanches, the aboriginal inhabitants of the Canaries, betrays discrimination after death. The various compartments accommodated important persons, while common people were buried outside the walls. From Estudio antropológico de los esqueletros inhumados . . . *by Miguel Fusté.*

dead are disposed of in a consummate manner. Whatever the merits of annihilation by fire, water, or ferocious beasts, architecture is the loser. For all we know, it might never have made a start, had not some paleolithic sexton hit at the idea of covering the dead with earth.

Interment is a different matter altogether. The compelling desire to recognize in death man's momentous entry into eternity rather than the termination of his journey on earth is as old as humanity itself. Neanderthal man, a subhuman being, who had not yet achieved an erect posture, buried the dead in carefully excavated trenches, and the better part of the world has not much improved on this expedient. Interment, with its attendant trappings and pomp, is common among the adherents of guilt religions, and men of business can be counted upon to make the most of it—architecture found in it its earliest application.

Burial does not necessarily entail disintegration of the body; some people like to keep in shape in the hereafter. In contrast to the wish for annihilation exists the desire for preserving the dead in as lifelike a manner as the art of the undertaker will permit. This desire is of long standing. Before embalmers began to avail themselves of formaldehyde, carbolic acid, and glycerin, they employed more appetizing substances for pickling, such as honey, butter, or salt. Honey was used for preserving Alexander the Great, a man reputed to have had such sweet body odor as to seem soaked in perfume. Salt usually went into the stuffing. The practice was looked upon as anything but odd; Old-Testamentary Joseph had his father embalmed in the heathenish Egyptian way which took forty days.

In man's quest for a simulacrum of lifelikeness in death, the macabre sometimes becomes indistinguishable from the maudlin. Who has not in some museum come across the mummy of a princess of the Middle Kingdom in the company of her mummified cats! However, it would be a mistake to equate human taxidermy with decadence gone awry. Nobody could accuse the ancient Aleutians—a branch of Eskimos—of having been tainted by Old World mentality; yet their mortuary practices nearly matched those of the ancient Egyptians, with the difference that in their egalitarian society everybody was entitled to the same treatment. Destined as they were by fate to an inordinately frugal life, they went overboard when it came to honor a leave-taking family member. The eviscerated body was filled with dry grasses, masked, and sometimes covered with a wooden armor—the poor man's version of the Chinese funerary jade suit of recent fame. A touch of Tussaud was added to these tombfooleries by setting up the dead in natural poses: men beating drums, women sewing, dressing hides, or nursing their infants. Moreover, the dead person was not turned out of the house but continued to occupy part of it, thus blurring the distinction between dwelling and tomb.[3] The

133. The Dwarfiestone on the Orkney island of Hoy is a memento of the mania for doing things the hard way. The block of sandstone, 28 feet long and 6½ feet high, was hollowed out to accommodate a tomb which, like so many tombs, later was converted into a habitation. A door is the only opening. (*Crown copyright, Royal Commission on Ancient Monuments, Scotland*)

usage was widespread, and the presence of human bones in prehistoric ruins often led excavators to believe that they had come upon a tomb when actually they had dug up a dwelling.

Home burial persisted into historic times. Houses containing tombs were discovered in Athens, a fact which induced Varro, the learned Roman, to trace the ancestor cult to this macabre manifestation of family solidarity. Among the aboriginal New Zealanders a man was buried in his own house with everything it contained. The doors were sealed and no one ever entered it again, with the result that in many villages half the houses belonged to the dead. A wasteful practice, it is by no means unique; among the Eskimos "when a person is evidently dying, they place with him everything which can soothe and comfort his last moments, and then leave the igloo, or house, which they close up, thus converting it into a tomb."[4] In countries where funerals are *de rigueur* and the churchyard is regarded as the legitimate resting place, capricious people, and none of the worst, prefer to bide their time in a private subterranean pied-à-terre. Self-indulgent Richard Wagner, for instance, in a mute assertion of sepulchral elitism, keeps distance from the crowd in afterlife by being buried on his home grounds, the garden of Villa Wahnfried.

While birth has always been, and still is, a more or less casual affair, death calls for intricate rites. Today, even a middle-class funeral is a major undertaking which leaves a big hole in one's posthumous pocket. Yet in the past, care of the dead was often far more demonstrative than in our time. Bodily comfort was seldom neglected. To put the dead at ease, his knees were slightly bent and an arm put under his head, as if he had dropped off to sleep. Or he was made to recline, preferably on a couch, a guest at an eternal banquet, with jars

134. The headstones of a Moroccan cemetery give a good imitation of gabled housefronts.

and drinking vessels at arm's length. Indeed, the wish to give the man who had everything in life no less in the beyond called for the equivalent of a dowry. A distinguished corpse was not put away without being supplied with clothes, headgear and all sorts of frippery; tools, furniture, toys, musical instruments, and a surplus of toilet articles. (Razor blades of archaic manufacture came in handy since facial hair does not immediately stop growing after death.) And weapons, let it be noted. Obviously, the doctrine of resurrection of the flesh is no less attractive to the primitive and heathen than it is to us. Occasionally, pet animals, wives, slaves, and hangers-on, alive or dead, were added for company and good measure. Although these mementos were of not much use to the deceased, they are the delight of tomb robbers and archaeologists.

Conspicuous display of sorrow is not just peculiar to the rich and great. In his peasanty ways a peasant is as keen on funerary etiquette as a pharaoh. He would feel guilty of disloyalty did he not lavish proof of his devotion upon his dead and, wherever burial has not been commercialized, the grave still has the connotation of a second home. Burial means but a change of residence. To uphold this fiction, a kind of disembodied cupboard love was cultivated in countries where food is highly respected. Hence grave goods sometimes include kitchen utensils and victuals. So far, however, food for the dead has not found its Escoffier.

This is not the place for enlarging on the composition of memorial meals or dishes prepared for the interred; a single recipe must suffice: In Bulgaria of bygone days, the dead were presented on the fortieth day after their departure with a cake made of boiled grain, sugar, almonds, parsley, and pomegranate seeds. Libations of wine were poured over the grave, while black coffee—an elixir of life in that part of the world—was served through an opening in the mound. In some countries, a pipe and a bottle of brandy complemented the repast. Clearly, our obituaries and requiems are no compensations for the solicitude which people of simple mind and simple habits lavish on their dead as a matter of course.

Death does not undo family ties. At least, it does not wherever mourners are bound by honor to extend the traditional courtesies to the dead. A flying visit to the cemetery, a prayer, a nosegay from the garden, a dab at a tear—such tributes of affection are fine for workaday occasions. Anniversaries, on the other hand, call for massive celebrations in exuberant spirits, with the graveyard as the appropriate setting. I remember how, many years ago, in Naples and points southeastward, on All Souls' Day entire clans convened for a picnic on top of their favorite forefather's grave, to feast and drink to his otherworldly health far into the night. Nowadays such respectful

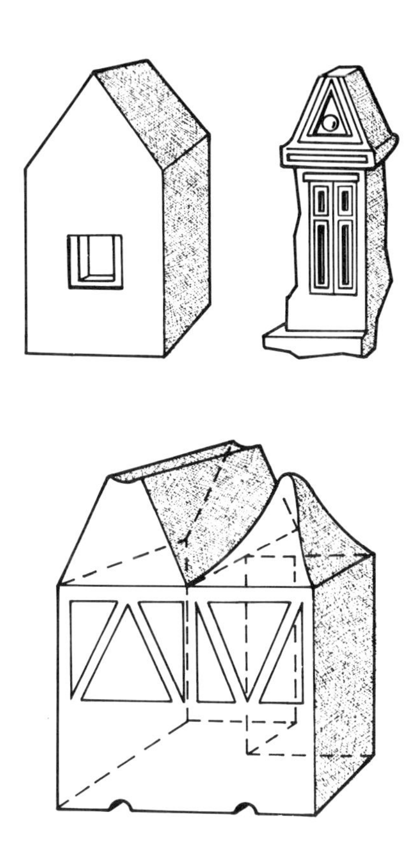

135. Stelae in the shape of houses were found in the necropolises of Celtic peoples. (After Linckenheld)

136. Funeral triclinium at a Roman cemetery for banquets in commemoration of the dead.

137. At the time of the Han dynasty (206 B.C.–A.D. 220), modeling a pigsty was not beneath a potter's dignity. (Courtesy Danish National Museums)

attentions may be in decline, bereaved families being content with half-listening to a canned mass for the dead on the radio.

The custom of funeral repasts in situ is of long standing. The Neapolitans, who retained more than a quirk of their distant forebears, merely went through a performance which in ancient times was known as *dies parentales*, when people repaired in procession to their family's tombs, to present oil, milk, and honey to the dead and to partake of a meal with them. (The offerings were poured into the grave through a funnel.) Again, the ritual did not originate with the Romans but had been adopted from Egypt, Assyria, and Babylonia. A habit so deeply rooted in people's minds as to survive into our time cannot be dismissed as farcical.

In the long run, though, ritual squanderings drained the mourners' sustenance, and ways had to be found to stop the wholesale transfer of household goods to the grave. The need for lopping off funeral expenses was met in the same mean spirit which prompts people to deposit wreaths of artificial flowers on tombs. By way of compromise the dead were presented with token gifts, that is, toy-size replicas of model houses.

This subterfuge bequeathed on posterity a knowledge of historic and prehistoric domestic architecture such as no written or pictorial records could. In Egypt and the Near East, in China and Japan, house urns of every conceivable shape have come to light. While some of them represent shrines, there is no dearth of ordinary dwellings. Graves have disgorged model collections ranging from huts, houses, houseboats, and wagon houses over granaries and stables to pigsties

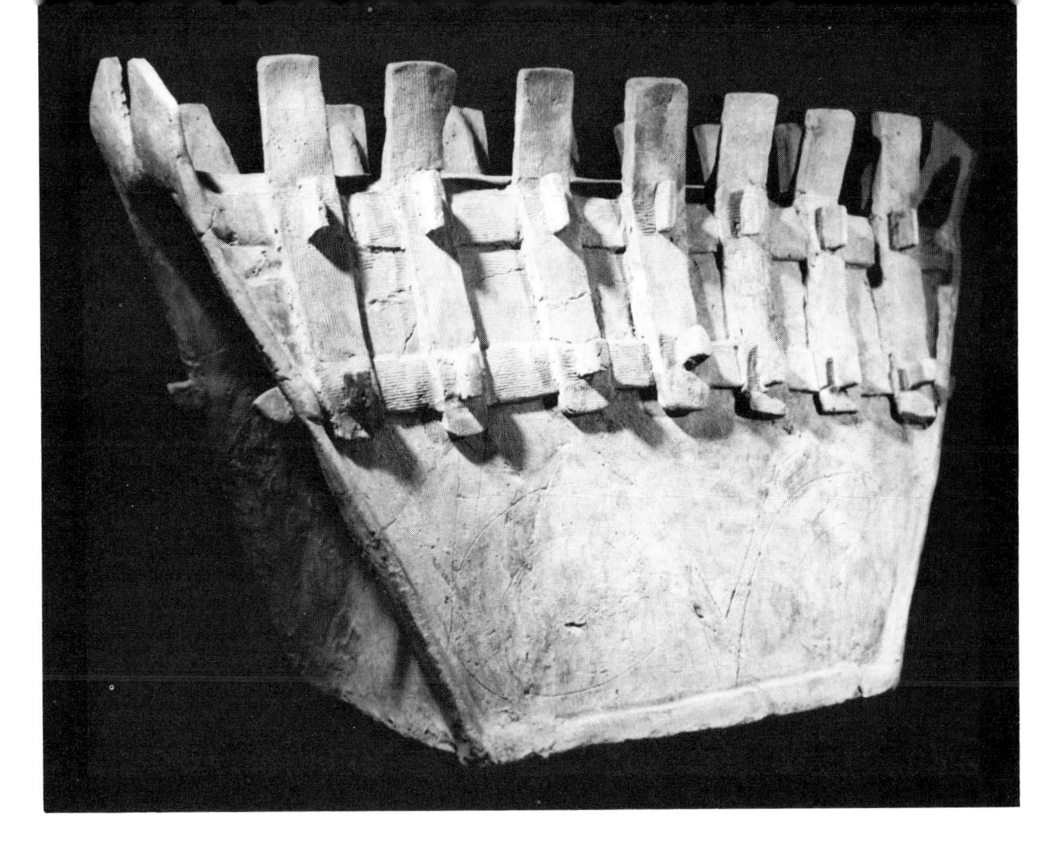

138. The best information on prehistoric Japanese dwellings and storehouses can be gleaned from haniwa, *realistic clay models intended as gifts for the dead. Their subject matter ranges from human figures, over utensils and weapons, to replicas of fifth-century* A.D. *architecture. Unearthed in Gumma Prefecture. (Courtesy André Bloc)*

139. Cinerary urn in the form of a pile dwelling, found in an Italic grave. Seventh century B.C. *Museo Etrusco Gregoriano, Rome.*

140. The absence of windows in a haniwa house model suggests its representing some kind of storehouse; its pillars might point to its use as a granary. (Courtesy André Bloc)

141. Opposite page: Haniwa house model found in Saitama Prefecture, Japan. (Courtesy André Bloc)

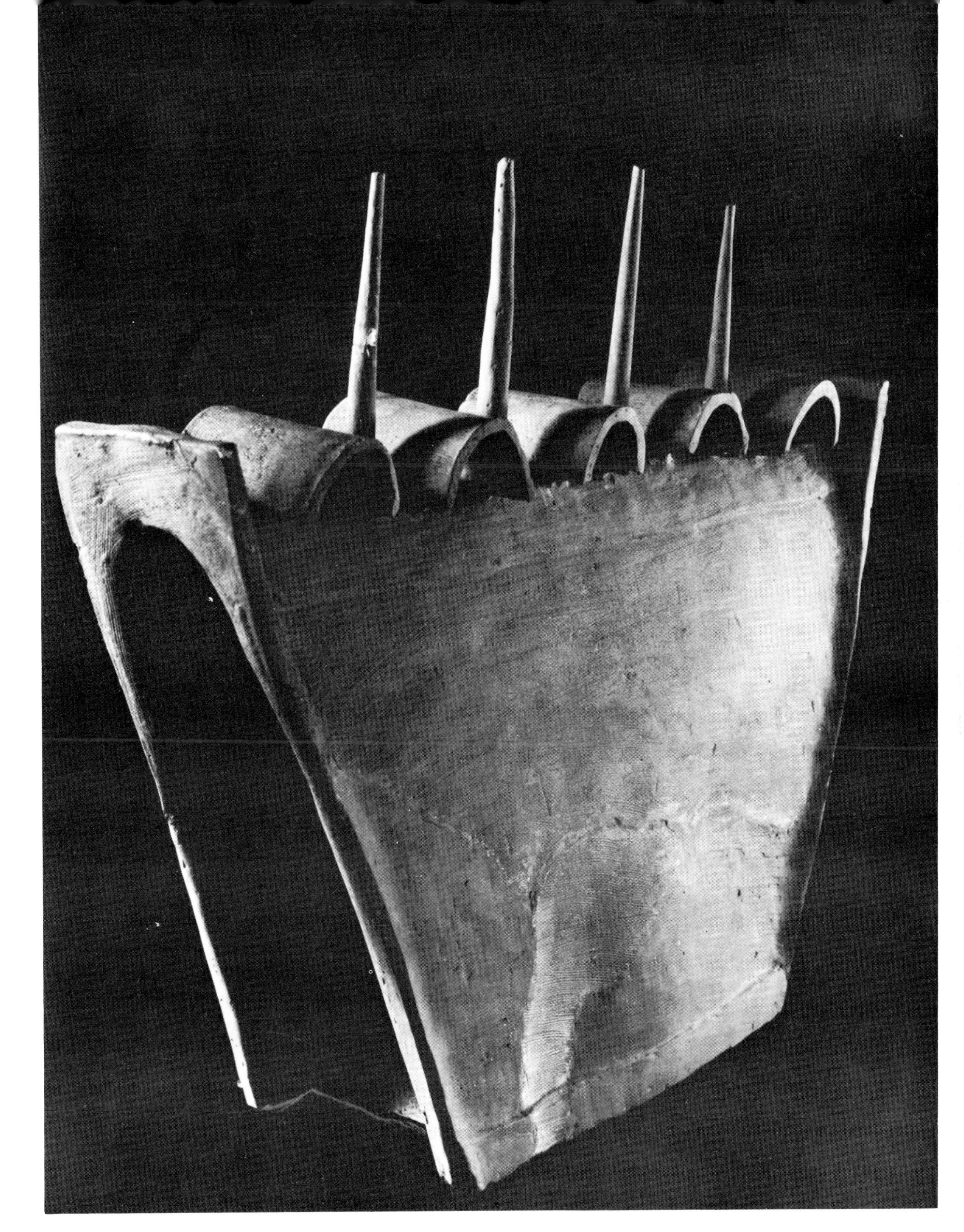

and privies. Many of these little effigies are executed with great accuracy, illustrating such details as frame work and roof construction, the manner a door was barred, or how a wall was painted. Since practically none of the portrayed dwellings survived—excepting some meager ruins on knee-high masonry stumps that leave everything to one's imagination—these miniatures are often the only clue to the vanished vernacular.

A wealth of information about the nature of early dwellings is supplied by a quite different type of pottery models that were discovered in Egyptian graveyards. They are not, as one might think, urns for human ashes but are purported to be houses for the souls of the deceased. Their manufacture has been dated—as much as Egypt's shaky chronology permits—from about the twenty-third to the eighteenth centuries. B.C. The earliest of them are shallow trays to be filled with offerings. Popular fancy identified the rim of the tray with a wall, while the space enclosed by it was thought to represent a courtyard. Perhaps such imagination is puerile, like that of the child who cuddles a stick for want of a doll; yet the consequences were gratify-

142, 143. Funeral votive trays in ancient Egypt included coarsely modeled houses, their courtyards crowded with household implements and, occasionally, domestic animals (opposite page). (Courtesy British Museum)

ing, for the would-be courtyard became the fertile ground for elaborate architectural improvisations.

"Probably all the features of the models," conjectured the abovementioned Petrie, "are copied from actual houses. They are the same as those in modern Egypt; the *satáh* is a roof enclosed by a dwarf wall, which is a usual safeguard on Oriental houses; the *mulgaf* is a hood to catch the wind and drive it down into the house; the *'eshsha* is a screen from the sun, supported on poles."[5] These models were playthings no more than our plaster madonnas; they were religious articles. Like food for the dead, they were not to be touched. They were meant to put mourners into the proper frame of mind. "It was the low-built rural house to which the affection of the bucolic Egyptians went out," writes archaeologist N. de Garis Davies; "it was a dwelling of this sort that he prayed for, and with a representation of it, therefore, he was wont to illustrate his prayers."[6] Unlike our

Neapolitan shopkeeper, the Egyptian's dreams were modest. Indeed, the two are miscast in their roles. For it is the Italian who indulges in pharaonic musings, the Egyptian who nursed a nostalgia for a simple cottage. The first probably never set eyes on any monuments other than the garish confections of his *campo santo*, while the latter could not have failed to fall under the spell of those monstrous burial heaps, the Pyramids. Nevertheless, his votive offerings make it clear that he cherished country-style living. Amusing trifles though the soul houses seem, they are invaluable documents, for they vividly portray rural and suburban dwellings of five millennia ago, give or take a thousand years.

144. Wooden model of an Egyptian granary circa 1800 B.C. *Grain is introduced through the three hatches. British Museum.*

145. Whether this model of a Minoan house, found on Crete, was intended as a funeral gift or as an architect's maquette for a client is irrelevant; it is remarkable for its timelessness. The well-proportioned openings, the variety of spaces obtained with few elements, are felicitously rendered in a rather unwielding material. Archaeological Museum, Iraklion, Crete.

146. Another view of the same house model.

Models of archaic dwellings also have been recognized in the Cretan clay sarcophagi called *larnakes*. (*Sarcophagus* means "flesh-eating," a ghoulish reference to the caustic properties of *stone* sarcophagi which consumed a body in forty days.) The noncarnivorous, Cretan kind—gaily painted terra-cotta coffers—do not impress one as macabre; the melancholy that muffles all burial paraphernalia here is absent. They resemble rustic cabins, dachas for the deceased, rather than coffins. According to one school of thought, they imitate huts, the oldest form of indigenous dwellings, slightly set off the ground, built with branches and twigs, and covered with a gabled roof. Others have pooh-poohed this interpretation. Archaeologists think it more likely that the larnakes served in the house as baths before being turned to sepulchral use. The theory is unattractive; to use a discarded tub for a coffin strikes one as the height of niggardliness, and the thought of rehearsing one's own funeral every time one takes a bath may have been unsettling to the frail.

147. *Just as houses sometimes served as tombs and vice versa, there also was no sharp distinction between sarcophagi and bathtubs. Archaeological Museum of Iraklion, Crete.*

148. This charmingly decorated twenty-two-hundred-year-old Cretan clay sarcophagus in the form of a hip-roofed hut makes one acutely aware of our coffins' wretchedness.

All these tubs, coffins, and house models were formed from clay, a material of such momentousness that it merits a detour.

The invention of earthenware goes back to the seventh day of Creation. After a week's hard work the Lord took a break and sought relaxation in modeling Adam's life-size figure from the clayey soil of the new planet. The question—What came first, pot or house?—is thus answered by the Scriptures: pottery preceded architecture.

The episode is by no means uniquely of Hebreo-Christian tradition, for other religions acknowledged the primacy of earthenware. The attribute of Ptah, the Egyptian god of primeval beginnings, is the potter's wheel on which he fashioned man. Prometheus, too, is credited with having molded the first man of clay, and so is Tahiti's chief deity, Taaroa. Similar myths are encountered in Africa, America, and Oceania. With minor variations the divine invention of the human prototype has served as a sort of plumb line for getting our bearings in this world. Although we sometimes become bored with our body, or irked by its built-in flaws—"We still have a trace of the Earth, which is distressing to bear," says Goethe's Faust—we have not been able to improve upon it. No matter; in our civilization Adam's prenatal effigy marks the birth of figurative art and the recognition of clay as ideal artist's material.

There is little doubt that in some parts of the world the pot preceded the house; pots were among the first objects shaped by man. For thousands of years he turned out vessels of exemplary beauty—if we arrogate ourselves to judge an art to which we are not equal. It is a sobering experience to walk out into the street from Iraklion's archaeological museum's Room I (neolithic pottery: 5000 to 2600 B.C.) and to come face to face with the pathetic contemporary trash exhibited in arts-and-crafts shops.

The size of the ancient pots is impressive. But then they were no ordinary pots. Some amounted to small silos for storing oil, wine, grain, dried fruit, and—the dead. A corpse, neatly folded, tied into a bundle, was stuffed into a pot which became his coffin.

Pottery is believed to have been invented in Iran in the seventh millennium B.C. When Iranian prehistoric man, a hoary type by any reckoning, still lived in caves, or at best in holes dug out of the ground and poorly roofed over with branches, he was already a fairly competent potter. Building a shelter of beaten earth may have suggested itself to him by his method of making pots, that is, by spreading the clay on a hard surface and beating or treading it into shape. Thus, although the origins of domestic architecture are many, in this case pottery seems to have been one of its godmothers.

149. The basket work of the Moroccan picnic hut admits light and air. (Courtesy Hispanic Society of America)

Old as the hills as pottery may be, it nevertheless has its precedent—*tace* theology—in the basket. One theory has it that the first pots were formed in a wooden mold lined with a mat from which the finished product was extracted and the mat peeled off. Another surmise traces pot making directly to the early custom of coating and lining baskets with clay and hardening them by fire. The pot emerged when the basket was consumed in the firing process.

Just as some granaries are outsize pots, many a hut is but an outsize basket. Wattle and daub, for instance, is essentially a basketwork technique in which the upright stakes are the warp, with withes woven in the weft and tied together with vines, and wet earth applied on this framework both inside and out. This type of construction was common in Britain in the Bronze Age, and although no traces of the actual buildings were found, we know what they looked like from ancient bas-reliefs. The wattle-and-daub technique survived in Europe until the eighteenth century and is still employed in some northern countries. Vitruvius sneered at it. An afficionado of rural buildings, he drew the line at the ungainly methods of barbarian tribes. "As for wattle and daub," he wrote, "I could wish that it had never been invented. The more it saves in time and gains in space, the greater and the more general the disaster it may cause; for it is made to catch fire, like torches."[7]

Earthenware pots and wicker baskets are still with us. Despite the advent of plastic bags and plastic bottles, or perhaps because of it, the tactile and visual appeal of the primitive containers remains undiminished. Originally intended as jugs and panniers, they ended up as houses.

150. Highrise pots. Sudan. From Auch im Lehmhaus lässt sich's leben *by René Gardi.*

151. Opposite page: From Morocco's Atlas Mountains all the way to Tunisia pastoral tribes keep their provisions in storehouses that resemble citadels. At Nalut in Lybia the local storage fortress is a vertical labyrinth, as confusing as Minotaur's den.

152. Above: Storage huts in southern Anatolia.

153. Right: Grass-roofed granary at Kindo, Ethiopia.

154. Among the most renowned architectural relics are the thirteenth-century B.C. *storerooms in the royal palace of Tiryns, Greece. The ceiling is formed by overlapping layers of projecting stones.*

Yet while baskets and pots found their way into museum vitrines, their offshoots, storage houses, are thought to be too inartistic, too prosaic, to be received into the pantheon of architectural history. To the professional aesthete there is little about them to warrant attention. They do not excite his curiosity. Only dead storage—tombs, catacombs, and columbaria—merit his sympathetic consideration. He could not be more mistaken, as the example of some storehouses will show.

In the past, food's quasi-sacral status was acknowledged by all civilized mankind, and the safekeeping of crops spawned granaries, an essential branch of domestic architecture. Pliny, discoursing on storing corn, noted in his *Natural History* that "some people recommend building elaborate granaries with brick walls a yard thick and not letting them admit draughts of air or have any windows. . . . In other places, on the contrary, they build granaries of wood and supported on pillars, preferring to let the air blow through them from all sides, and even from below."[8]

Galicia would be one of the "other places." Her native *horreos* fit Pliny's description, except that they are built of stone. Their outstanding characteristic is a substructure of mushroomlike pillars. The reason for the wide-brimmed capitals is not far to seek; they are rat guards. The height of the columns is probably justified on aesthetic grounds unless Swiss rodents are clumsier than Spanish ones (see figure at right).

Some of these earthy granaries come within an inch of formal architecture. Severely geometrical and of meticulous workmanship, they exude the devotional atmosphere of an ecclesiastical edifice.

155. Rat guards similar to those of Iberian granaries top the supporting pillars of huts near Zermatt, Switzerland.

156. Galician granaries rival the sumptuousness of royal tombs. Although crosses and urns confer upon them a distinctly cemeterial flavor, they are thought of as lifesavers. The cross calls to mind the ancient concept of the holiness of food, the multiplication of the bread, and the host of the Holy Communion.

To confuse matters further, they also exhibit an affinity with certain Roman funerary monuments. With or without crosses, they are the very image, if considerably larger, of sarcophagi hoisted up, as it were, and left in abeyance, the pallbearers having stopped in their tracks and turned into stone pillars (fig. 157).

157. Sarcophagus on stone pillars in the necropolis of Hierapolis in Asia Minor.

Whether these petrified barns came out of antiquity's dustbin or whether they were independently invented by the local peasantry is a moot question. They probably have been around in more or less the same shape since primeval times. Prehistoric cinerary urns representing similar granaries which came to light in northern Europe are likewise rectangular and covered with a roof of double slope. They, too, rest on pillars with rat guards.

Corn cribs like the one in figure 158 make part of the common stock of the Portuguese vernacular. A mutant of the Galician strain—or is it the other way around?—it is well contrived and well made by any norms. Yet even a cursory glance reveals that it was put up without recourse to plans, sections, and elevations, indeed, without a thought to measurements. The stone and wooden posts are not spaced with the regularity of an edifice plotted on the drafting board; they fell into place as work progressed. Typical though such license is for the vernacular, it finds no convincing explanation. Why do peasants who trace furrows in the fields with utmost precision have no use for precision in their buildings? Exact spacing of pillars does not require higher learning. What makes them disregard the most elementary conventions of professional builders? Surely, it could not have escaped them that the wooden posts on the third floor of this granary are far from equidistant; that in the fourth bay there is a swing from duple to triple time.

To relax our prejudices and to better assess the conflict between nonconformity and pedantry we would have to set up a conspiratorial control test—say, erect a hypothetic Greek temple as perceived by untutored men of pastoral background. Well, let's do it.

For the sake of speculation, and to avoid philistinism toward the teachings of architectural history, let us imagine an archetypical sort of building, its walls girded by columns loosely laid out, their intervals determined by the erring eye rather than by the measuring cord. All capitals vary in their dimensions and proportions. Moreover, while most of the columns are of stone, a few are shaped from timber or built up with bricks.

158. Opposite page: Portuguese granary from Arquitectura popular em Portugal. *(Courtesy Sindacato Nacional dos Arquitectos, Lisbon)*

The pious will have to brace themselves to conceive such architectural heresy, but a dash of erudition might help to regain their composure. For here, archaeology comes pat to the occasion. It so happens that the most ancient temple in Greece, the Heraeon at Olympia, was exactly the unruly type we have conjured up. It had

none of the obsessive blandness that one associates with Periclean Parnassianism. A shaggy kind of building, supremely authentic, blatantly archaic, it had wooden posts instead of stone columns. As these rotted, one after another, they were replaced by stone columns. However, this took a long time—eight centuries, to be precise. When Pausanias, that indefatigable traveler, visited Olympia in the second century after Christ, he could see the last of the wooden posts. A whiff of the sacred grove still clung to the temple.

Yet even the stone columns showed the most marked differences. They were as dissimilar as trees, their diameter varying from 39 to 50 inches. Their fluting was not identical. Of the nineteen capitals that were found, no two are alike. In short, the earliest of Greek temples were devoid of that unrelenting precision that we are asked to admire in Athens's or Nashville's Parthenon. They reflect the unconstraint of a rural society that successfully made the transition from worshipping among trees to worshipping in a cell surrounded by stylized tree trunks.

Another instance of the derivative nature of formal architecture, and also of the kinship between granaries and graves, is provided by Lycian tombs. The Lycians, who lived in a coastal district of southern Asia Minor five hundred years B.C., buried their dead in tombs cut into perpendicular rock. "The tombs of Myra," writes archaeologist Machteld J. Mellink, "give us a literal and three dimensional illustration of Lycian timber architecture. In the sixth to fourth centuries B.C., the Lycians created stone replicas of some of their timber building types, presumably age-old forms, now made visible in the careful details of rock façades, rock cut chambers and sarcophagi."[9]

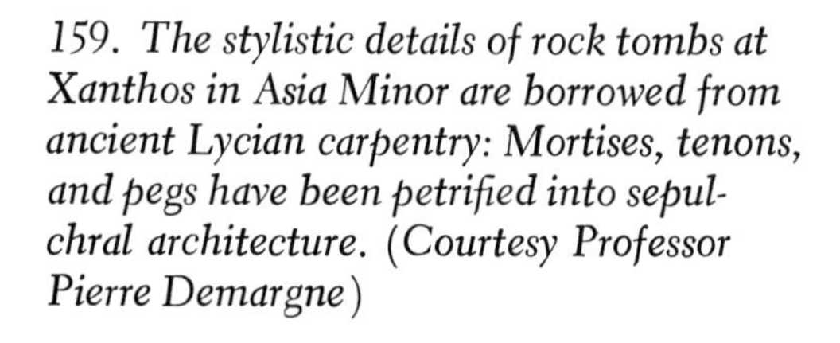

159. The stylistic details of rock tombs at Xanthos in Asia Minor are borrowed from ancient Lycian carpentry: Mortises, tenons, and pegs have been petrified into sepulchral architecture. (Courtesy Professor Pierre Demargne)

160. Nineteenth-century granary (with its modern roof omitted) in the plain of Elmali, Turkey, where the local type of carpentry has not materially changed since the Bronze Age. (Courtesy Professor Machteld J. Mellink)

Since the local peasants can hardly be expected to harbor antiquarian interests in these tombs, it is all the more surprising that, due to them, many features of Lycian carpentry have come down to our days. "Not all types of houses and huts are nowadays made of timber. The main building form with ancient characteristics is that of the granary, whether of small size for the average family supply, or large for the wealthy landowners," explains Professor Mellink; "a survival of some architectural forms through five millennia is not improbable in the Lycian area."[10] The nineteenth-century granary which she adduces as evidence proves that the millennia have not vitiated the onetime woody robustness.

Strongholds

Among the few places on earth that haven't suffered from the blight of organized tourism and are still safe from the scheming of travel agents is the "Garden of the Holy Virgin," the more-than-one-thousand-year-old republic of Athos at the dead end of the homonymous peninsula in northwestern Greece. So far its isolation from the profligate world has been about perfect. In compliance with an imperial edict dating from A.D. 1060, it is strictly off limits to children, males under age, and, paradoxically (considering its exalted Protectress), to all females, including animals.

In other words, Athos is infinitely remote—in time if not in space—representing as it does a singular anachronism, a semiutopian social order based on medieval concepts. Its trifling population, whose number has dwindled from forty thousand in the sixteenth century to a few hundred, and the low density of construction would make it the most desirable place to live in, this side of Heaven, were it not that its inhabitants are a mirthless and misogynous band of men, ill disposed toward life's pleasures. They are monks who thrive on prayer and little else.

Lack of roads and a riven mountainscape that attains a height of more than six thousand feet discourage hiking. Besides, Athos is covered with near-virgin forest of cedars, cypresses, sycamores, giant oaks, pines, and sweet chestnut trees, abetted by an impenetrable brushwood of laurel, wild roses, myrtle, and any number of odorific evergreen plants. A nineteenth-century visitor to the mountain,

161. Opposite page: The monastery of Simopetra, one of twenty monkish fortresses on Mount Athos in northern Greece, rises from a green wilderness crossed by mule paths only.

R. Curzon, thought the scenery "so superlatively grand and beautiful that it is useless to attempt any description."[1] And it is true that its admirers grope in vain for hyperbole and metaphor in their struggle to do it justice verbally.

Athos's architecture, or what is left of it, is in no way inferior to its natural assets. Of the 180 monasteries that are said to have dotted the mountain in the eleventh century, 20 are still standing. No two of them are alike. The many-storied Simopetra—to single out a building of distinguished mien—has the appearance of a lamasery or, if you will, of a faintly sinister castle, albeit without a donjon, dungeon, and moat. For lack of a drawbridge, a fiendish contraption, the hoisting basket, has been in use until a few years ago. It is a primitive elevator, without pushbuttons or motor, for the vertical transportation of all sorts of things, including people. Made of withes, it nevertheless looks more like a cage than a basket. Three factors assure a safe rise: solid wickerwork, a strong rope, and the right muscle tone in the bicepses of the monk who turns the wheel. Receiving the passengers

162. "Then the disciples took him by night, and let him down by the wall in a basket" (Acts 9:25). *He is Saint Paul, escaping from prison in a basket on a rope, the forerunner of our elevators. Eighteenth-century Flemish tapestry.* (*Courtesy Victoria and Albert Museum*)

one by one, he has ample time for inspecting and, if need be, rejecting them. Although would-be pilgrims today are screened in up-to-date bureaucratic rites far from the Holy Mountain, the basket has not been retired altogether. Easy to maintain, independent of mechanical power and its failures, it now serves common household needs.

Even though Simopetra ranks as one of Athos's most sturdy buildings, a visit to it is not recommended to anyone unpracticed on the flying trapeze. A close brush with eternity awaits one at the guest quarters, located on the top floor of the outermost wing. The sagging wooden galleries that gird the cyclopean walls, which in places provide the only communication between the rooms, do not just look fragile; they are. A single misstep will cut short your journey. The floorboards, as thin as shingles, curl up under one's weight, while the *trompe-l'oeil* railings stand upright by sheer inertia. Self-preservation has taught the monks a way of floating angellike over these aerial corridors—no doubt a matter of faith rather than gravity, like walking over glowing coals. Seeing you blanch, your guide hastens to tell you by way of encouragement the story of a fellow monk who, while carrying a tray with coffee cups along the very gallery you are standing on, broke through the treacherous planks and fell several hundred feet to his death. And here one might suppose the story ends. But no; the doomed monk shortly emerged from a crack in the rocks, hale and unperturbed, the glasses on his tray in place and not a drop of coffee spilled. At Simopetra it pays to take out afterlife insurance.

Although these days the monastery's gate stays open from sunrise to sundown, the hospitality dispensed within is at its nadir; the unsuspecting visitor may find the night's lodgings consisting of plain wooden boards bereft of mattresses, sheets, and blankets, with only two diminutive pillows to cushion an ear and a hip. Fortunately, such penury is the exception; elsewhere guest accommodations measure up to the comfort found at youth hostels or police stations. The chief vexation is food of a sort that the poorest of poor might have eaten in a medieval town in the second year of a siege. To discourage the tourists from staying, he is offered little more than bread, warm water, and an unsavory pea soup in unattractive surroundings.

Simopetra represents much more than a petrified prayer; it is the nonpareil parable of architecture. If one can read between the stones, so to speak, the motives, noble or mean, that drive men to isolate themselves from the world here are built into the grand structure: Immoderate desire for solitude; a refusal to face those insults and humiliations inflicted upon them by their fellow creatures that we call, hypocritically, life's realities; but mostly an excess of misanthropy —all converge into an irrepressible want for a refuge that is able to provide something amounting to emotional amnesia. No philosophy,

163. A monument to misanthropy and piety, Simopetra ranks first among the most extravagant constructions on Athos. The fragile wooden galleries that interconnect the monks' quarters form a striking contrast to the colossal substructure.

no religion can bestow onto the tortured soul the comfort that is experienced from being surrounded by meter-thick walls. The prisoner immobilized in a dungeon may think otherwise, but then the difference between imposed and self-inflicted deprivation of liberty is slight; the palpable circumstances are much the same.

At Simopetra the danger of a bombardment, or siege, was far too remote to justify the cyclopean pedestal for a monastery of modest size. In its millennarian existence, Athos's republic was never seriously threatened by war; its ecclesiastical fortresses were conceived by sheer exuberance. Whether the monks were misguided in their zeal, or whether they did the right thing, those who do not share their faith cannot judge. Moreover, any attempts at a sober evaluation of this singular monkish stronghold are frustrated by the realization that it is a recent building. It is not at all a hoary monument but a late addition to the republic. Although it looks archaic, Simopetra was built in 1895. Evidently, the passage of a thousand years did not affect Athos's vernacular. Within the monasteries' walls a primitive sort of Byzantinism, made fast by the laws of the religious orders, always kept its rightful place. Chapels and churches fairly choke of decoration. The profane parts, on the other hand, the living quarters, the entire concept of building, are timeless and remained unaffected by the architectural vogues of the outside world.

Like convents, castles, the powerful metaphor of man's presumed self-sufficiency, are an Old World institution; the combination of residence and fortress occurs mainly in Europe, Asia, and Africa. A gulf of roughly ten centuries separates modern man from the time when castles were a common commodity, indeed, a necessity. Apart from ruins, what passes for castles today are the products of architectural taxidermy—mummified mansions, national monuments, and occasionally, houses of royalty (itself a moribund species). Although Americans are not averse to anachronisms and have produced their quota of pseudomedieval churches and colleges, they seem to draw the line at castles. Castles are as un-American as a mistress; the man who has everything collects credit cards, not castles. To him a castle is not a house or a home but the moldy container of the darkest aspects of the Dark Ages. Incapacitated for feelings of grandeur by his upbringing, he has as much difficulty imagining himself in feudal surroundings as wearing armor to his office. To come across an inhabited medieval castle is as much of a shock to him as hearing a museum's dinosaur neigh.

Castles are the fossils of architecture, an uncomfortable reminder of the time when wars were still fought with arrows and lances; when killing was a handicraft rather than an industry. No architect within

164. *Colonialism is unabashedly expressed in this juxtaposition of garrison architecture and indigenous dwellings. The presence of the Danish Christianborg Castle on the Guinea coast reduced the adjacent African village to something like an architectural eczema that had to be stayed. Detail of an eighteenth-century engraving. (Courtesy Danish National Museums)*

living memory has been commissioned to build a castle, and I don't mean a château. West Point, Alcatraz, and San Simeon—all of them brave attempts to graft heroic silhouettes onto the landscape—received no plaudits. (Despite its forbiddingness, Washington's Pentagon is not a latter-day squashed castle but the apotheosis of barracks architecture.) Ruskin confessed that he could never, "even for a couple of months, live in a country so miserable as to possess no castles."

Neither do urban armories or those now-defunct town houses, optimistically referred to as châteaux, rate as castles. The word *château* is the same as *castillo*, but the stiletto-sharp syllables of the latter have been blunted and the *s* has given way to the broad circumflex. Châteaux lack the grimness of their Castilian counterparts. Indeed, some of the messier ones seem to have been designed by pastry cooks. Popular belief relegates quite rightly cream puff châteaux to flat country, along lakes and rivers. Could anybody conceive of a château's siege except perhaps in terms of an operetta, with a corps de ballet costumed as amazons, hurtling peaches and melons to the strains of an effervescent battle music?

Nathaniel Hawthorne, far from being the product of a country with a feudal past, instinctively understood the basic topographical

exigencies of a true-blue castle. Chillon, that perfect picture postcard château, with snow-covered mountains in the background and snow-white swans in the foreground, its "snow-white battlements" (Byron) reflecting in the Lake of Geneva, "is woefully in need of a pedestal," he maintained; "if its site were elevated to a height equal to its own it would make a far better appearance. As it is now, it looks, so to speak, profanely of what poetry has consecrated, when seen from the water, or along the shore of the lake, very like an old whitewashed factory or a mill."[2]

Today, when we are trying to rediscover the forgotten ties between architecture and the natural environment, we ought to take a good look at castles, those pivotal points in the landscape. Topographical accents par excellance, they are always commensurate to Nature's

165. In the age of chivalry every crag seems to have extended an irresistible invitation to cap it with a castle. The Arabic seaport of Aden was protected by a veritable picket line of citadels perched on the rim of an extinct volcano. From Georg Braun's Civitas orbis terrarum *(1581).*

scale. In particular, that loftiest place of abode, the crisp, rock-bound castle, recommends itself to our attention. An excrescence of the rocks, it perches on a hill or mountain in acrobatic suspense. It borrows its protective coloration from the environment in much the same way that a practiced Japanese "steals" the trees' silhouette of his neighbor's garden. It is all stone; no leprous stucco, no peeling paint. The feudal touch is usually provided by an eye-catching cliff, a mesa, or a dense forest, all of them anathema to the plebeian. Wherever *he* puts up a building—in the country or in town—he is bent on annihilating its natural setting. Any outcrop—a tree, a boulder—must be pulverized so that he can better visualize his creation. He is a leveler at heart. For his flight of thought he needs, as it were, an airstrip, an expanse of nothing.

166. Weary of see-through architecture, some of today's architects build pure volumes, their surfaces unimpaired by any openings. The Portuguese frontier fortress of Marvão, its only ornament the rich texture of stonework, anticipates such hermetic architecture.

Fleckenstein Castle, the aristocrat's idea of a proper domicile, is but one of a cluster of fifteen castle ruins, some within shouting distance of each other, in the Alsatian county of Lembach, near Strasbourg. Admittedly, the 1589 engraving by Daniel Speckle typifies the sort of pictorial reporting that strains one's credulity. Although the castle's contour seems familiar from some turn-of-the-century American hotels, the general appearance is frankly feudal. The audacity of concept and construction is the more remarkable as the castle occupies a base only 7 yards wide. Speckle, who was an expert on the subject—he wrote a book on fortifications and masterminded Gibraltar's defenses—readily conceded that "one cannot understand how such a work came into being." Nor can we. But neither should we dismiss Fleckenstein as a chimera.

Speckle knew his way around what he called the wondrous house. "A magnificent merry dwelling it is," he wrote, "with stables and cellars cut into a rock that rises like a four-cornered diamond, straight as a wall."[3] He failed to tell how the inhabitants and animals reached

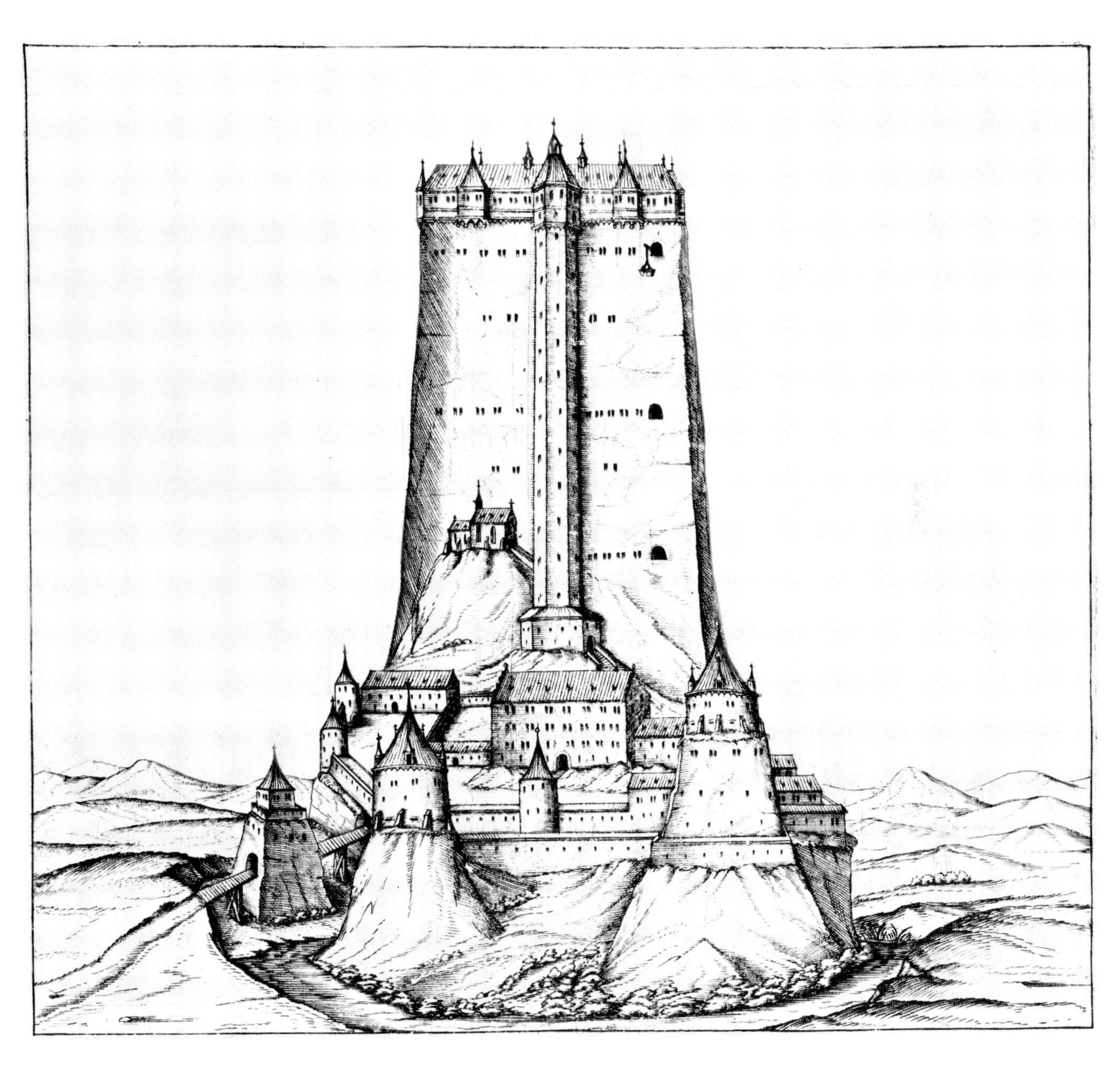

167. An organic skyscraper, so to speak, the amalgam of a most improbable rock formation and an unremarkable building, the Alsatian Fleckenstein *(* ∘ *eleventh century,* † *1680) was one of ten thousand castles that once dotted* German *territory. (Courtesy* Spencer Collection, New York Public Library*)*

168. Loarre Castle, a typical eleventh-century fortification in the Spanish province of Huesca, was built on the site of Roman ruins. The extensive remains, which measure a third of a mile in circumference, are classified as a national monument. (Courtesy Evelyn Hofer)

the top, whether by the slender lean-to staircase—did it contain a ramp to be climbed on horseback?—or by riding the primitive elevator whose cabin, a basket, can be seen in the engraving. Surely, the energy expended in negotiating this protoskyscraper was out of proportion to the tininess of its premises.

Believed to be inaccessible and inexpugnable when built in the eleventh century, Fleckenstein became progressively less so during the six centuries of its existence. Several times restored and improved, it burned down in 1680, a casualty of modern warfare. All that's left is a 170-foot stump of the former rock.

169. The earliest of man's fortifications did not much differ from the snowy ramparts piled up by elks as defenses against predatory wolves.

In the beginning a castle was just an earth mound surrounded by a dry ditch. The mound's crest may have carried a palisade no more protective than the one the Pilgrims erected around Plimoth Town. Indeed, about the year 1000 the art of building defenses in Europe had not developed beyond the know-how the Assyrians brought to constructing the walls and towers of Nineveh. It was only during the Crusades—a span of almost two centuries—that the incombustible castle emerged. Out of this unspeakably hideous chapter of history came a better understanding of architecture in general and of fortifications in particular. Discounting treachery, a castle built on high ground often remained impregnable; that is, until gunpowder came into use. It was perhaps the only time when a walled-in and well-provisioned populace was justified in experiencing a feeling of absolute security in the face of an invader.

But life was not thought worth living in a castle that was never assailed by forces stronger than the elements. Only during a siege did its true merits come to the fore. The battlements afforded a grand view of men locked in gladiatorial combat—a Colosseum turned inside out, with the grandstand in the center and the stage in the

170. Scandinavian combustible wooden castles conjure up knights on rocking-horses. The 1502 drawing by a lansquenet shows Älvsberg, a bone of contention between Danes and Swedes.

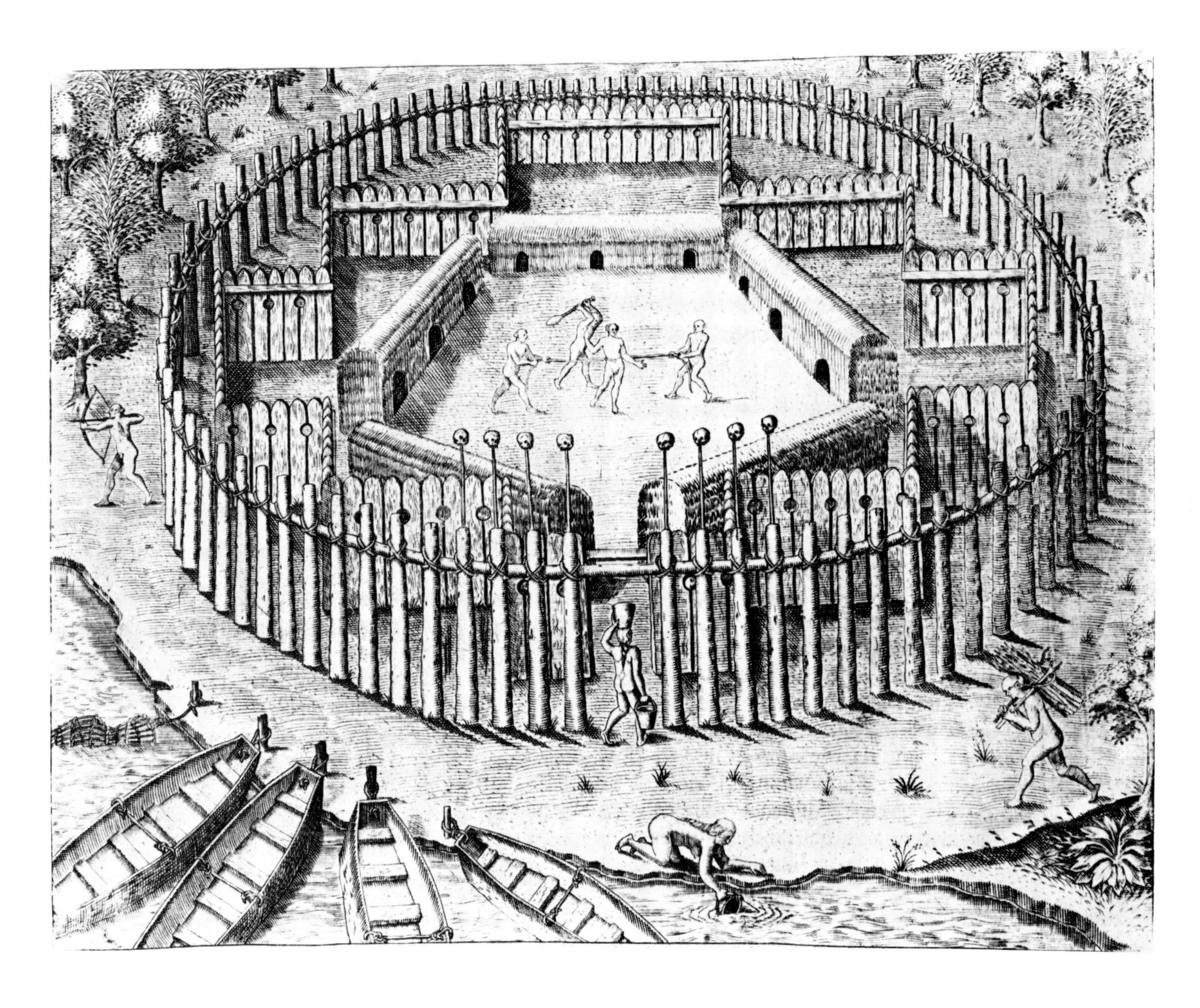

171. The wooden stockade, an improvement on that most elementary type of fortification the thorn hedge, was used wherever timber was plentiful. It equally served Brazilian cannibals and the Pilgrims at Plimoth Plantation. The engraving is from Abelin's Historia antipodum *(1631).*

round. Even when the attackers succeeded in breaching a wall, little was lost. The defenders just hopped from one perch to another, dousing assailants with pitch and molten lead, missiles as deadly as bullets. A siege in full swing pinpointed the imperfections in layout and structure, and the burgrave would make a mental note to have the level of the moat—a castle's Plimsoll line—slightly raised, or to make the hard-edge towers round. Castles so strong that their bastions were never put to the test remained inviolate until wreckers began to quarry them for building materials.

172. The demolition of a castle was as tedious and expensive an undertaking as its construction. Artillery, usually beyond the means of the warring parties, had to be rented. Fifteenth-century woodcut from the Schwäbische Chronik. *(Courtesy Prints Division, New York Public Library)*

Donjon, moat, and ramparts a castle do not make. The archetype of castle, the lofty nest of a knightly bird of prey, can perfectly do without them; of all architectural species it is least subject to building conventions. To ferret out castle architecture in its natural habitat one must go to Spain, where one is rarely out of sight of castles or their ruins. The northern highland, peppered with peels and keeps—20-foot-thick walls hugging a 2-foot staircase—is known as Castile, the name being derived from *castillos*, the frontier defenses that once guarded the country against the Moors. (Similarly, Austria's easternmost province is called Burgenland, literally, castle country.) Today, when the sites of a good many castles have been "developed," "landscaped," or otherwise defaced, one must seek the true aerie to which no road, no path, ascends. Usually, it is partly in ruins, superbly aged, bathed by the rains and swept by the winds, those self-appointed curators of castle architecture. The aesthetic pedant who alights on the castle pitched high on a mountaintop is dismayed by what he takes to be its unruliness. Its very substance proves to be elusive because it seldom permits one to optically grasp its exact shape. One simply cannot walk around it; an inspection of its circumference might lead to a broken neck.

There are compensations. Only up there does one get the perfect bird-of-prey's-eye view of the world by now as remote as the stars—a scented landscape covered with groves of gnarled olive trees whose trunks resemble Chinese ideographs; a modest river, icebound in winter, swollen in spring, dry in summer, when veritable oleander woods emblazon the stony bed; a narrow lane, a cloud of dust kicked up by an ambulant herd of sheep; a camel-back bridge; a walled cemetery; and, low in the castle's shade, a village of whitewashed houses. It is a setting like this that fixates what Berenson called the aesthetic moment, "that flitting instant, so brief as to be almost timeless, when the spectator . . . ceases to be his ordinary self, and the picture or building, statue, landscape, or aesthetic actuality is no longer outside himself. The two become one entity; time and space are abolished and the spectator is possessed by one awareness."[4]

Orthodox architectural history deals for the most part with build-

173. Vélez-Blanco in the Spanish province of Almeria is dominated by a polygonal sixteenth-century castle.

174. The shape of medieval feudal towers anticipated such twentieth-century bastions as New York's World Trade Center. San Gimignano, Tuscany.

175. Opposite page: In the city-states of medieval Italy, amid the hothouse growth of ecclesiastical and princely architecture, stand some tall weeds, the towers of the feuding nobility who used to settle their disputes in aerial combat high above the plebs. Bologna's Garisenda tower, leaning menacingly over its neighbors, dates from the year 1100.

ings that have served mankind as physical and spiritual first-aid stations—temples, cathedrals, theaters, and thermae—to the exclusion of permanent habitations. No doubt the towering castle is somewhat unwieldy, too intricate to be reduced to plans and façades. Yet castle anatomy is fairly simple. Although no two castles are quite alike, they can be broken down into a few basic patterns. In the publications of the International Castle Research Institute the castellologist Count de Caboga defined three "genuine" types: the square plan, which he calls the Arabic-Byzantine type; the Ringburg, or round, plan; and the irregularly shaped castle, which closely follows the terrain. Nearly all are built on the principle of the enclosed courtyard, the noble feature of the han, caravanserai, and Roman villa. The bulk of the buildings usually consists of precise geometrical volumes—cubes, cylinders, cones, and their subtle modulations—so close to the vision of early-twentieth-century architects, the will-o'-the-wisp called functionalism. No towerlets, balconies, or stony curlicues detract from the overall tidiness. (Crenellations, we must remember, are not ornaments; every tooth hid a man intent on pouring death onto the assailant below.)

Taking into account the enormous diffusion of castles in former times—medieval Germany alone possessed more than ten thousand of them—the vast number of their inhabitants, and the fact that a country's population amounted to about one-tenth of what it is today, castles ranked high as dwelling places. Indeed, it would be a mistake to see them as purely military installations. They belong more nearly in the category of civilian shelter that in wartime gave refuge to peripheral folk, villagers, peasants, even burghers of neighboring towns. These emergency uses are all but forgotten; the picture of the castle that takes hold of our imagination in childhood is that of an architectural toy tenanted by a fairy-tale prince with the looks of a page boy—an insipid creature, a youth, not a man. Not even in the hirsute era of Freud and Marx did he wear as much as a mustache. He never burned down villages or tortured his enemies. He did not give battle except when slaying dragons holding decorous, latter-day Andromedas in their clutches.

The bona fide knight was on intimate terms with the supernatural because his lair seemed to exert a special attraction for spirits, good and evil. But then, the earth was still flat, the heavens were hanging low, and divine intervention was more frequent than in our days. Saints still took a personal interest in people, and so did the angels. To assist their charges, guardian angels would dive from a rift in the clouds to a faultless landing on a castle's battlements. Moreover, every castle had one or more resident ghosts as a matter of course, and was regularly visited by devils and demons. Best known in this respect is the Wartburg in Thuringia, the scene of the devil's encounter with

Luther, who, displeased with his visitor, threw his inkpot at him. (He missed. The ink was of the best quality, for the spot it left on the wall can still be seen.)

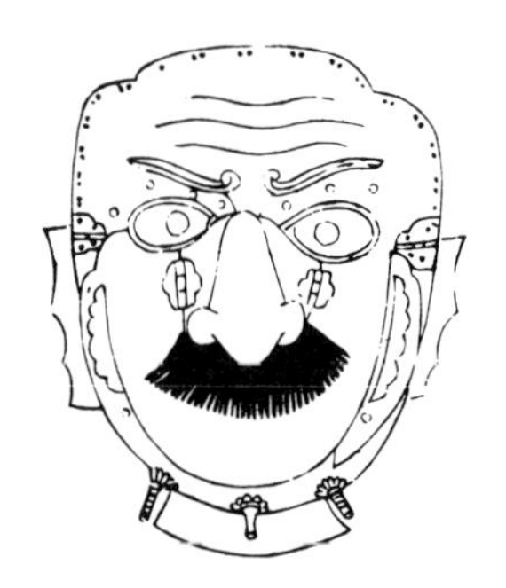

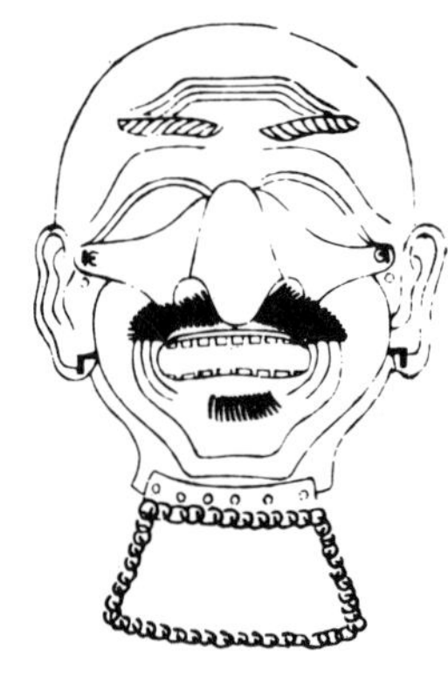

176. Iron masks from a seventeenth-century reference book on Japanese armor.

A short etymological note on knighthood is perhaps in order. Americans, who skipped the more interesting stages of Western civilization's rise, and entered it only in its last, decadent phase, look at things feudal with the kind of distrust, not to say disbelief, that stems from a total lack of experience. Hence, to put flesh on the skeleton of nonpedigreed castle architecture, it would seem proper to introduce at some length its inhabitant, the knight, a species as dead as the aurochs.

Originally, the term *knight* applied to a person of modest station in life; the Old English *cniht—Knecht* in German—signified lowly helper. Romance languages, forever more polite and to the point, stress his superior means of transportation. The words *chevalier, caballero, cavaliere* clearly refer to his equestrian status. Time has eroded this meaning; in today's Spain every man is a *caballero*, as a glance at any toilet door will confirm.

Some of the more exalted knights have been transmogrified into demigods inhabiting Valhallalike castles whose dimensions remain unsurpassed by modern engineering. The classical example that helped to make mythomania so attractive is Arthur's Camelot. Thanks to Broadway and Hollywood, Artus de Bretagne is as familiar to our generation as Batman. In fact, some heroes of the medieval epic poems behave quite like comic strip characters. In the *Chanson de Roland* four knights kill four thousand Saracens; Roland himself continues fighting with his head split open. Yet Arthur (who was not a king but a general) becomes more elusive the deeper one gets entangled in the particulars of his household. Just take a sober look at the magical Round Table, a piece of furniture so grand in conception and execution that most architecture looks puny by comparison:

It seems that in olden times problems of precedence were as pressing as they are today. For lack of a proper protocol Arthur's knights, an intensely social set—although, like French gourmets, they admitted no women at table—were forever at each other's throats to advance their standing, or more accurately, their seating. To put an end to their quarrels, tells the thirteenth-century poet Layamon, a carpenter built a portable table that seated sixteen hundred men. (There is no reason to think the poet was inventing, we are assured by the *Encyclopaedia Britannica.*) Now, Ramsay and Sleeper's *Architectural Graphic Standards*, the modern builder's book of etiquette, allots to each person 2′–0″ for comfortable seating. Unless the knights sat in one another's laps like stacking chairs, the table's diameter came to more than 1,000 feet, leaving several acres of waste space in the center. No known castle, nor, for that matter, any present-day structure, could accommodate such an outsize piece of furniture.

The *historical* knight appears less mysterious. Not overexigent when it came to life's necessities, he was, however, avid of certain pleasures, such as hunting man and beast. Masculine to a fault, he was brimming with fighting spirit and, like the udder of the lost mountain goat, itching to be relieved. He possessed courage in excess of loyalty; indeed, he was implacably mercenary. El Cid, Spain's greatest hero, thought nothing of switching sides, free-lancing for Christians and Moors alike. It seems that a knight was bound only loosely by the code of his profession, if profession one can call it. He was probably far more practical than one might suppose, hard-fisted and stiff-necked, if only by virtue of his shiny battle outfit. Since armor, his second line of defense, is a key to the understanding of castle architecture, a word about it would seem appropriate.

177. *The windowless walls of feudal castles have their counterpart in the knight's hermetic business suit.*

Knights are the Crustacea of the human species. Their workaday clothes may have been rustic, if not altogether tawdry, but their formal dress was not. A coat of mail was not just a survival kit but an extension of the castle, much as a car is an extension of the house. Like a car, it needed greasing and polishing and the smoothing of dents incurred in collisions and accidents. The services of a first-rate armorer were therefore a knight's best life insurance and a boost to his *amour-propre*. Any red-nosed, grizzly-bearded knight, clapped into his portable fortress, was the picture of perfection. For armor is far more symmetrical than the man inside. Although stylized, it is a good counterfeit of the human form down to the tiniest bumps: the pigeony chest, coquettish waist, knuckles and fingertips, slitted eyes in a reptilian face (no mongolism, merely a defensive squint), flaring nostrils but no metal ears. An armored knight's merciless adversary was the sun, which unavoidably, turned his steely suit into a

178. Fighting was probably suspended on dog days when knights, roasting to a crisp under their tight covers, found umbrellas too cumbersome in hand-to-hand combat. Sixteenth-century drawing from a tailor's manual.

broiler. Umbrellas, introduced in Renaissance days, not only marred his martial appearance but badly impaired his mobility.

Knights fighting one another clashed like cymbals, but the clangor of battle was inaudible to them thanks to their splendid acoustical isolation. Nevertheless, armor was the thing; the men merely provided the stuffing, a sort of inner tube. Since armor is extremely rigid and far more body-building than a backboard, it must have been difficult to ascertain in the heat of a skirmish whether a galloping knight, solidly anchored in his stirrups and propped up by his scaffolding, was in control of his faculties, or badly wounded, or stone-dead. An unhinged, *fallen* knight was worth several times his weight in scrap metal. Patched and hammered into a second-hand suit, his armor would serve yet another knight or knave.

Another clue to understanding the knight and his castle is the food consumed in feudal times. We do not know what was served at Arthur's table. We have, however, a detailed account of the victuals consumed (centuries later, in 1578) on the occasion of Wilhelm von Rosenberg's wedding at Krummenau in Bohemia, now Český Krumlov. Among the ingredients of the meal, or meals, are listed 370 oxen, 113 stags, 98 wild boars, 2,292 hares, 3,910 partridges, 22,678 thrushes, 12,887 chickens, and 3,000 capons. These meaty viands were balanced by proportionate amounts of eel, pike, carp, and salmon, and 5 tons of oysters. In addition, 40,837 eggs were broken, and 6,405 pails of wine were dispensed to maintain appetite and promote digestion.[5] Pantries resembled city markets; kitchens took on the aspect of blast furnaces.

Nor was the entertainment accompanying the meal of the puny kind we are used to today. Instead of somniferous luncheon speeches, meals were served to the dulcet sounds of flute and shawm. There were mock combats, riding and dancing exhibitions, fireworks, and mummeries. At Rosenberg's wedding a touch of the circus was provided by monkeys riding on goatback while playing the harp, a feat that Hollywood could not duplicate for money or good words.[6]

Of course, such gorgeous fare was the exception; poor knights often had to make do with potluck and to listen to the high winds plucking chords from the fir trees. In the noblest castles the sustenance consisted of crows, cranes, ravens, storks, swans, and even vultures, each and all muscular birds, low in gustatory lure but high in allegorical content. In anticulinary England meals were usually of the beggarly kind: salted fish, tough legumes, and indigestible desserts were the order of the day. The entire cooking for the week was done on Sunday, to be warmed up at 10 A.M. for lunch and at the ungodly hour of 4 P.M. for dinner.

It would be unjust to gauge the former amenities of castles by our own scale of values. People who rave about 12-foot ceilings in brownstone houses might find breakfasting in a six-story-high baronial hall a transporting experience, available today at small expense in those castles that have been converted into hostelries. Most of us, though, have never developed a taste for spaciousness, cooped up in mini-apartments as we are most of the time. Another insight into the knightly luxury is afforded by the description of their beds. Some were of heroic size, and if the mattresses did not always correspond to our idea of molluscan comfort, the bedspreads were sometimes precious beyond our commercially conditioned imagination. What modern housewife could conceive of bedcovers made of ermine and black sable, and of veritable Oriental tents for canopies? The Styrian minnesinger Ulrich von Lichtenstein described his ladylove's voluptuous bed as conscientiously as our home-furnishing editors. Her velvet mattress was covered by two silk sheets, a comforter, and blissful ("wunniglich") cushions. At the foot of the bed stood two candelabras, and the walls of the room "were hung with a hundred lights."[7]

Medieval castles were the last repositories of antiquity's cultural know-how and of those Oriental technical accomplishments that crusaders gleaned in their unholy wars. To mention but two: The fifteenth-century Guadamur Castle near Toledo was equipped with a hypocaust, the floor heating developed by the ancient Romans. Another Spanish castle, the eleventh-century Peñafiel in the province of León, had ventilation shafts reaching down to the underground storerooms. With the ruination of the castles vanished also the knowledge of their amenities. The empty shells, hideouts of owls and urubus, do not let one suspect that, in some instances, they once were the wardens of the most advanced dwelling culture of their time.

The flowering of castles produced some exotic blooms far from Europe's *Burgen* and *castillos* on the secluded islands of Japan. In that avalanche of books on things Japanese that was triggered off at the end of the Pacific War, intelligence on native architecture was all but smothered by the sultry sentiments of its rediscoverers. The shrines and palaces won the palm, with teahouses coming in a close second, yet, perhaps because martial fervor had dropped to an all-time low, Japan's castles remained largely unnoticed.

In Japan, the art of fortifications was, if not introduced, at least greatly improved by, of all people, European missionaries. Like Japanese armor. Japanese castles have a flavor all of their own. The cyclopean walls that rise from the moat, of an elegance unknown in the Western world, form a strange contrast to the highly flammable wooden structures inside. For sheer size, few other nations can match Japan's building stones; the largest of them, used in the construction

179. Opposite page: Beginning a day in a six-story-high breakfast room greatly predisposes one to savor architecture on a grand scale. This festive interior is part of the former castle of Emperor Charles V at Fuenterrabía, Spain, now run as a government inn.

180. Himeji Castle near Kobe is considered the acme of Japanese feudal castles. In a complete about-face of the nation's cultural vandalism of a century ago, this building complex was belatedly declared a national treasure. (Courtesy Chanticleer Press)

of Osaka Castle, is 19′–2″ high and 47′–6″ long, or about the width of two brownstone houses and the height of two floors. The keep always surprises by its exquisite carpentry and downright homelike atmosphere. Above all, it is as spotless as an old-fashioned inn. Naturally, one has to take off one's shoes before entering it, a ritual that cannot have but a pacifying effect on the most sanguinary of soldiers. Can anybody imagine an American general leaving his shoes in front of the Pentagon?

About a century ago, when the Japanese went all out to embrace Western-style progress, the sight of their antiquated castles made them uncomfortably aware of their backwardness. Sensing in their architectural heritage a hindrance to the emancipation from their medieval past, they simply decided to do away with their castles. They leveled the bastions and filled the moats and publicly auctioned off the take-home parts. A turret went for the equivalent of three American dollars, a donjon for ten, while the wooden structures of an entire castle, sold for fuel, brought the sum of seventy-five dollars.[8] Castles that remained standing because they found no ready takers were incinerated in the raids of World War II.

Since then the Japanese had second thoughts about their destructiveness. In the 1950s the reconstruction of Osaka Castle sparked off a castle boom that is long in subsiding. The feudal strongholds that patriotic zeal destroyed are now recreated as bulwarks of tourism. (Between April 1959 and March 1960, 7 million people visited Osaka Castle.) Chambers of commerce are rivaling historical societies in resurrecting the past, and castles are rising from the ashes in their old splendor with not a few innovations. Thus, the woodwork now is hung from steel and reinforced-concrete frames, and elevators comfort the lame and footsore. Today, Japanese castles are built as local attractions, and by no means for foreigners only, in places where there had never been a castle before.

In Europe castle architecture meets with similar rehabilitation. Even to ruins comes momentary glory by being submitted to the nocturnal treatment conferred these days to the Acropolis. Plunged into a dirty-orange glow and electronic sound, they provide paltry entertainment for the man in the street for whom theater and opera are forbiddingly highbrow. Although in Europe castles are rarely put up for sale, they have a collector's value, the collectors being not individuals but well-endowed institutions or the state. Since 1926 the Spanish government has been slowly but persistently converting castles into state-owned hotels. This is not a newfangled idea; when chivalry ceased to rule, and knighthood dissolved in the acid test of time, a number of nobles took up the lucrative business of catering to the lowborn. They never parted with the insignia of their former

181. These prisonlike towers denote, paradoxically, a luxurious hotel; the thirteenth-century Moorish fortress that dominates the Spanish town of Jaén was skillfully restored and turned into a state inn.

status, and turned their coats of arms into signboards. All those White Horses, Black Eagles, and Red Lions one encounters so frequently in central Europe ostensibly hint at the innkeeper's ancestry.

Each year tourism's vanguard captures one more string of castles. For a nominal fee one has the run of palatial premises that no modern hostelry can match: stony edifices with walls from 3 to 10 feet thick, honeycombed with apartments—a paradoxical combination of vastness and downright Oriental economy of space. Each room is different in shape and size, differently furnished and decorated. (The coffered or painted ceiling is somewhat out of reach, but may be studied with the aid of binoculars.) Labyrinthine corridors and staircases lead to arcaded courtyards, patios, cool promenades, and gardens complete with arbors and fountains, a chapel, a belvedere, and semisecret nooks tucked away in towers and battlements.

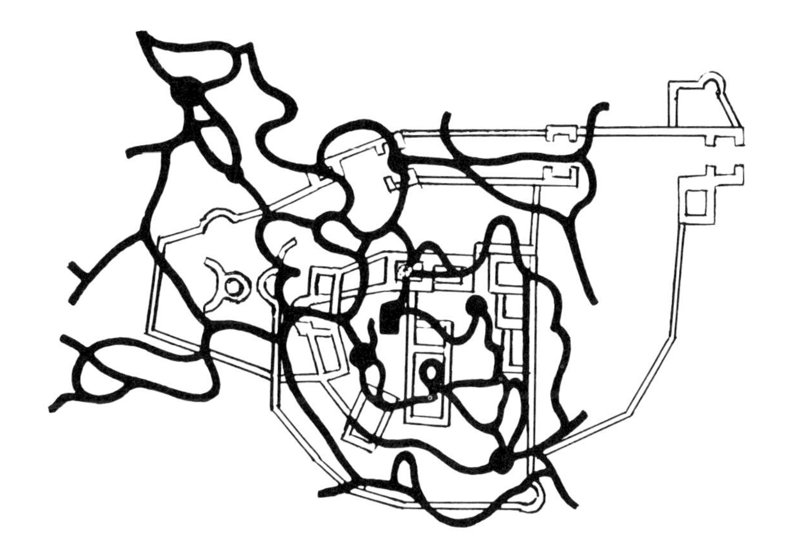

182. The underground passages of medieval castles often came as close to the labyrinth as its mythical predecessors. The conqueror of a castle, entrenched within its walls, sometimes was taken by surprise by the enemy who had stayed behind for months in hidden cellars, a sort of built-in Trojan Horse. The whirling lines, superimposed on the plan of Austria's Pottenstein Castle, represent secret corridors. From Piper's Burgenkunde.

Alas, the food is disappointing. One scrutinizes the menu in vain for wild boar, peacock, heron, or like heraldic birds; the fare is mercilessly up-to-date. Lately, though, efforts have been made at elevating the sights of the sandwich-and-ice-cream-conditioned tourist by staging "medieval suppers" at a number of castle inns. With drums beating and trumpets sounding, boards that groan under the load of food, or so it seems, are carried by pages in stately procession to the refectory. The table setting is discretely backdated with the help of porringers, tankards and goblets, wine jars, and water pitchers for washing one's hands. To keep the gastric juices flowing, a perambulating troubadour sings of bold men's bloody combatings and gentle ladies' tears, and of a feudal age's main preoccupation, the wars against the infidels. The dishes are worth a detour for their unfamiliarity: garlicky appetizers; beans fried with eggs; *sopa boba*, "crazy soup," mentioned in *Don Quixote*, a broth made from gizzards and pounded almonds; unholy concoctions of sugar, bacon, dates, and flour; or eggs fried in honey, which the nuns used to serve visiting worthies.

The discriminating eater, however, will do well to cross the border. Portugal counts among its *pousadas* a few reconstituted castles whose kitchens are singularly capable of pleasing one's palate; the parade of hors d'oeuvres alone has all the gaiety of a minor folk festival. However, the country that might triumph in a national contest to revive the medieval pleasures of the table is France. The very word *château* suggests rich food and vintage wine. A *château aux pommes frites*—short for Châteaubriand—or a bottle of Château Margaux, is, every bite and drop, artistically on a level with the creations of, say, Viollet-le-Duc, the supreme restaurateur and confectioner of medieval architecture.

183. Until the early decades of the twentieth century the villages of Svanetia, a mountainous region in the Caucasus, bristled with defense towers. Every house was equipped with a donjon, intended as temporary shelter in troubled times.

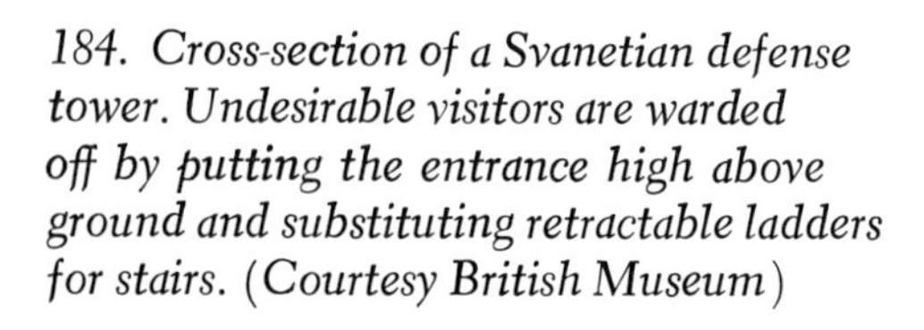

184. Cross-section of a Svanetian defense tower. Undesirable visitors are warded off by putting the entrance high above ground and substituting retractable ladders for stairs. (Courtesy British Museum)

From textbooks of history and architecture one might infer that erecting citadels was the sole privilege of the aristocracy. To correct this impression, those least-known manifestations of peasant genius, Transylvanian village fortresses, are introduced herewith.

Transylvania itself is something of a nonplace on most people's map of the world. At best the name brings to mind horror-filled castles of Dracula fame, or early cinematic operettas set in a Ruritanian never-never land. Transylvania, which means, literally, "beyond the woods" (of Hungary), lies in today's Rumania and was known for the past eight hundred years as Siebenbürgen, so named after seven fortified towns founded by Germans. It is a fertile high plateau, about half the size of New York State, where three-quarters of the population are engaged in agriculture. In the twelfth century the country was colonized by Saxons who, three hundred years earlier, had been preceded by a Magyar tribe, the Széklers—sacrificial mini-nations that soon turned oppressors of the local Vlachs. The Saxon settlers were peasants from the Lower Rhine, Maas, and Moselle, jealous custodians of their heritage, and singularly intent on preserving their ethnic identity. To the Saxons and Széklers—who formed self-governing communities—fell the arduous task of defending the eastern border of the Hungarian kingdom against the incursions of

185. The grim ramparts of the triple-walled village fortress of Harman in Rumania stand for an epitaph to man's faith in walls as protection against the brunt of warfare.

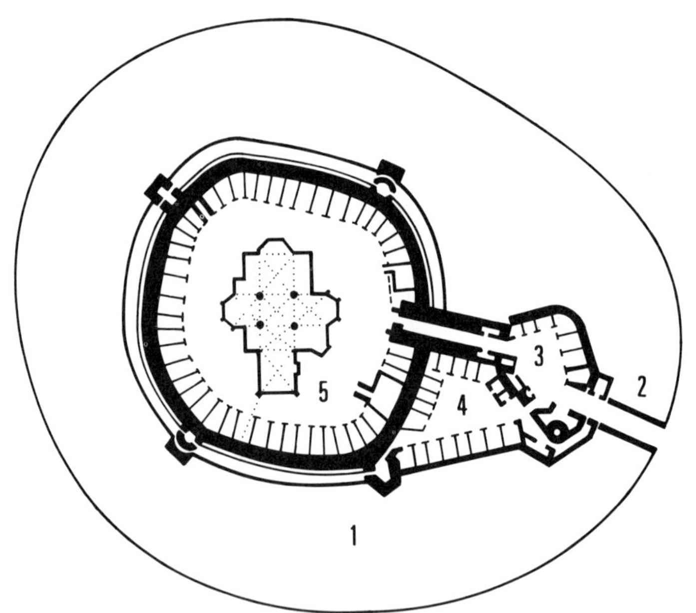

186. Right: Plan of the heavily fortified churchyard at Prejmer (Tartlau), Rumania: (1) moat; (2) bridge; (3) forecourt; (4) baker's yard; (5) church.

187. Wartime shelter never looked more attractive than at the Transylvanian village fortress of Tartlau, alias Prejmer. Tacked onto the protective wall are the living quarters that provided temporary refuge during a siege. Below them are the stables and storerooms. To fully appreciate this communal architecture the mind's eye has to fill it (and the banalized churchyard) with an ebullient crowd.

Note: The two photographic fragments do not convey the vastness of the courtyard; the (undistinguished) church in the courtyard's center (see plan, fig. 186)—which hides a large slice of the lodgings—is not shown in the picture but only indicated by a buttress on each side.

188. At Harman, civilian wartime shelter clings to the church for spiritual and structural support.

Asian nomads. In return they were granted vast privileges which assured them territorial and political autonomy. Thus, far from being trodden-down serfs, they represented an elite whose pride and solidarity proved decisive in their long struggle for survival.

Their first test came with the Mongol invasions in the mid–thirteenth century when their towns and villages were razed to the ground. Galvanized back into life, they rebuilt the province and, as a defensive measure, strengthened their only refuges, the churches. By surrounding them with deep moats and wooden palisades—soon to be replaced by solid walls studded with towers as many as ten stories high—they arrived at their own peculiar kind of military architecture. However, on no account should these walled churches be confounded with fortified churches proper, those hermetic houses of worship that dot Europe from Scandinavia to the Mediterranean and also crop up in Mexico—monolithic fortifications, albeit without bastions, moats, and drawbridges. The Transylvanian churches themselves were not fortified; their defense was assigned to peripheral buildings.

In 1420, when the Turks under Muhammed I were overrunning the country, the Saxons were well prepared. By then they had built some three hundred parochial fortresses, each self-contained and capable of absorbing a town-size population. Two hundred of them are still standing. Those that have been honed to their former brightness are, despite a certain peasanty awkwardness, singularly handsome. A good example, though by far not the grandest, is Prejmer, the former Tartlau, 20 kilometers northeast of Kronstadt, alias Brašov.

The present buildings, which date from the thirteenth to the seventeenth century, consist of formidable walls, 16 feet thick and over 40 feet high. Onto the inner surface of the walls are grafted three tiers of defensive galleries with embrasures, firing slits, and spouts for pouring boiling water and pitch over the assailants. What distinguishes this alfresco architecture from fortifications the world over is the provision of living quarters for civilians. Accessible from the galleries are hundreds of wooden chambers, connected by stairs and ramps.

Like feudal castles, these village fortresses were built with long sieges in mind. At Tartlau there were enough stables to admit an ark's contents and cellars for food to last forever. Grain was ground in a horse-driven mill. Arms and munitions were stored in four towers of which two remain. If we are to believe local historians, in wartime intramural life went on as normally as the constricted space permitted. Children did not miss a day in school, artisans switched to war industries, and the peasants, deprived as they were of their fields, played soldiers. Although the defenders often had to battle an enemy tenfold in number, they repulsed every assault.

Close to Prejmer lies Harman, the Saxon Honigsberg, its church encircled by no less than three concentric ramparts. Here, people's cubicles cling like barnacles to the church itself. Admittedly, the Transylvanian churches lack architectural distinction, but then their function as sentinels of European civilization has been mostly symbolic. This is not to say that they are destitute of ecclesiastical art; they all have their quota of altarpieces, statuary, and organs. Yet the perils shared in the past made the parishioners overly practical-minded. Harman, for instance, possesses some good fifteenth-century frescoes in a side chapel which recently doubled as a storage room for lard.

The vanishing vernacular

An African villager looking for the first time in his life at a European house does not suspect the travail and anguish that go into building it—the ritual of buying the land with the help or hindrance of agents, lawyers, and local authorities; securing a bank loan or mortgage; preparing plans, estimates, and documents indispensable for the construction of the house; and paying taxes and insurance policies attached to it forever after. To him the result may look elementary.

Similarly, a Westerner inspecting an indigenous African dwelling may find it, too, quite plain. For what he perceives is but the tangible substance, endearing in its unpretentiousness, while the all-pervading magic escapes him. He may see in it the container of a life of extreme artlessness—or what strikes *him* as artlessness—and may envy the owner his freedom to build, untroubled by the chicanes of bureaucracy. Little does he suspect the intricacies that go into the house's planning, for there is a good deal more to it than meets the eye. Primitive people have a knack for complicating everyday matters to make one's head swim. The homestead of the Dogon, a tribe who live south of Timbuctu in Mali, will serve to illustrate the point. Without the painstaking inquiries of ethnologists we would never know that a Dogon so-called Big House is not at all a big house; it represents a man lying on his side, procreating. But let us dissect the Big House limb by limb.

It comprises a central room, the *dembere*, or room of the belly, around which are placed a kitchen, three storerooms, and a stable, and the big room whose entrance is flanked by four conical towers. "The plan of the building," write Marcel Griaule and Germaine Dieterlen, "is said to represent, on one hand, *nommo* (the son of God) in his human form, the towers being his limbs; on the other hand, the kitchen and stable are said to be the heavenly placenta and its earthly counterpart, together representing the head and legs of a man lying on his right side, whose other limbs also have their architectural counterparts; the kitchen represents the head, whose eyes are the stones of the hearth; the trunk is symbolized by the *dembere*, the belly by the other room, the arms by the two irregular lines of storerooms, the breasts by two jars of water placed at the entrance to the central room. Finally, the sex organ is the entry which leads by a narrow passage to the workroom, where the jars of water and the grinding stones are kept. On these, young fresh ears of new corn are crushed, yielding liquid which is associated with the male seminal fluid and is carried to the left-hand of the entry and poured out on the shrine of the ancestors."[1] We certainly miss a lot of uplift, sexual and otherwise, living as we do, according to the renting agent's lingo, in ordinary rms, kit, clsts, etc., and we are the poorer for it. Whatever the deeper significance of the Dogon architectural anthropomorphism, it endows a home with an inner life all of its own.

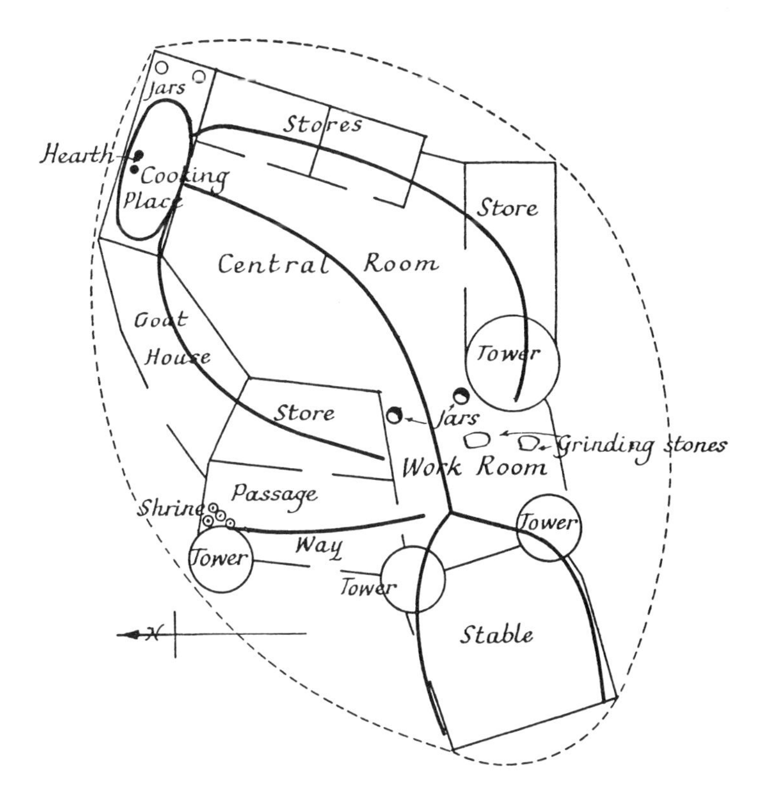

189. The diagramatic plan of a Dogon family homestead gives no inkling of the heavy-handed symbolism which pervades its every nook. It took anthropologists to disclose the cosmic significance of what seem but plain household furnishings. From "The Dogon" by Marcel Griaule and Germaine Dieterlen, in African Worlds. *(Courtesy International African Institute)*

Freud anticipated the anthropologists' interpretation by many years. Forgoing any sallies into the darkest Africa, he arrived at his own, starkly anatomical concept of what a house—any house—stands for. To him it was "a substitute for the mother's womb, the first lodging, for which in all likelihood man longs, and in which he was safe and felt at ease."[2] The doctor's contemporary compatriots may very well have cherished recollections of their erstwhile uterine accommodations but these days, given a choice, they probably will opt for the substitute.

For zoomorphic examples of domestic architecture: The Kiwai on New Guinea see their dwelling as a gigantic wild boar. On the Gulf of Papua, "the largest houses are built to represent an alligator with open mouth: The platform of the house is the lower jaw, and the long shade over the platform the upper, so that standing on the platform you stand in the alligator's mouth, the house sloping to appear as a body."[3] Americans should be able to muster a sympathetic understanding for this architectural menagerie, the more so as they have a near-equivalent, by now past its palmy days. The roadside vernacular of hot dog and milkshake dispensaries occasionally takes a turn at animal shapes—not of savage beasts but of the Donald Ducks and Mickey Mice of nursery fame.

If one had to single out the basic difference between formal and

190. The Galician farmhouse represents agglutinative architecture; each generation does its bit. The result is a house with a face as easily distinguishable as a human one. The warts are stones that hold down the slate roof. Province of Lugo, Spain. (Courtesy Hispanic Society of America)

vernacular architecture, it would not be sumptuousness versus artlessness. What distinguishes the two from each other are different sets of idiosyncrasies. Formal architecture is, with few exceptions, connected in our mind with symmetry, the vernacular with its absence. Symmetry is implicit in the concept of every noble building; it is the birthright of pedigreed architecture.

One only has to peruse art history for corroboration. Could anybody conceive of listing Egyptian pyramids or a lopsided Parthenon? Surely, an asymmetrical Versailles might have struck our forefathers as *lèse majesté*, an asymmetrical cathedral as blasphemy. If there are asymmetrical palaces and churches, they are so by circumstance, not by intent. Ever since architecture came of age, the abstract attributes of power and faith have been expressed by the mirror image.

Asymmetry spells uncouthness; at best it is associated with quaintness, at worst with dissoluteness. It may very well be that this want of visually expressed discipline puts the houses of peasants and fishermen beyond the pale. Of course, one does occasionally find in vernacular architecture both lateral and concentric symmetry, but on the whole it is accidental. As a rule, the houses are formed and deformed by successive additions of volumes quite unrelated to each other, a drawn-out process that is the very opposite of the professional builder's habit to make definite plans and see them through.

191. The vernacular is much more than a style; it is a code of good manners that has no parallel in the modern urban world. To judge from its looks, this Japanese farmhouse on the island of Shikoku could easily be a sacred shrine.

Architects look with a jaundiced eye at rural towns and villages; usually, it is artists and writers who are sensible of their charm. Landscape painters—a species earmarked for extinction—sometimes discern beauty and strength in ostensible chaos. The fascination of Calabrian towns, noted Edward Lear in his 1852 travel journal, "appear to consist in the utter irregularity of their design, the houses being built on, under, and among, separate masses of rock, as if it had been intended to make them look as much like natural bits of scenery as possible."[4] Such approval of the erratic may seem heresy to town planners who ply their trade with T-square and compass.

Obviously, spontaneous architecture is not to be judged by academic standards; what looks chaotic to the biased often has been knitted together into an obdurate sort of harmoniousness. All peasants have style, declared the great critic Sainte-Beuve. Nevertheless, peasant houses do not epitomize any critic's thesis; while they are not altogether deficient of a certain grandeur, they are never solemn. Above all, they do not lack identity, an essential ingredient for dwellings that aspire to dignity. At any rate, they are never of that menacing baldness which is the hallmark of most drafting board architec-

193. A walk through the Calabrian town of Pentedáttilo in places comes close to mountain climbing.

192. Opposite page: A crucial factor in all building activities is the choice of the site for a town, a village, or a house. A nervous, deeply lined landscape, even when bereft of vegetation, will attract people who are sensible to their surroundings. The hills which dominate the Spanish town of Calatayud are an example in point; towers and walls, the ruin of an eighth-century castle, and, for a rural touch, the threshing floors in front of the houses have tamed (and ennobled) what was to begin with a wild tract of countryside. From España, pueblos, y paisajes *by José Ortiz Echagüe.*

194. Casares, one of many similar Andalusian towns, is built in a truly native style—as distinct from an intellectually created form language. Its people never discuss architecture; indeed, they lack the appropriate vocabulary for arguing its merits and demerits. Nevertheless, they betray a nearly unerring instinct for the right shape and color and, most importantly, an uncanny feeling for harmoniousness.

ture. The venacular is much more than a style; it is a code of good manners that has no parallel in the urban world.

Vernacular architecture ranges so far in space and time as to defy summarizing. Most of it refuses to be parceled out into neat categories. To compound this intractability, vernacular and prehistoric buildings sometimes appear to be much the same thing, as when a present-day house can be dated back and compared to its prototype of an undetermined period. In the following are described a few bona fide architectural fossils that made the transition to the rural vernacular without missing a beat.

195. Opposite page: Until recently, domestic architecture in the Cycladic Islands was austerely geometric. Only a chapel or a clock tower was thought to merit some stylistic embellishment.

Up to our time islands were perfect safekeeping places for local specialties, including dwelling types. One particularly attractive house form survives, insecurely and disguised, on Thera, the southernmost island of the Cyclades. Thera, one of the strangest places this side of Heaven—the un-Greek climax of a Greek voyage, it is called—has been in the news since mythological times. It figures in Greek legend, it made history and prehistory. Through the centuries it has been enriched and impoverished by the occupation of Phoenicians, Dorians, Spartans, Franks, and Turks. Epigraphers, archaeologists, geologists, and vulcanologists have mined it for precious information. The apple of many a scientist's eye, the island has attracted and puzzled explorers of all times. To quote at random—in 1835, Ludwig Ross, professor of archaeology on the newly founded University of Athens

196. A zigzag path climbs the sheer cliffs of colorful rock to nonpareil Phira, capital of the Aegean island of Thera. Blindingly white houses and sustaining walls cling to the rim of a volcanic crater 700 feet above the sea.

197. Despite an elementary building technique, Thera's native architecture luxuriates in sculptural improvisations. Perpetual whitewashing of walls, roofs, and pavements lends to houses their unearthly gleam.

and keeper of Greek antiquities, having just arrived at Thera, thus tried to put his excitement into words in a letter to a friend: "I am writing to you, as it were, from a new world, a world of the most extraordinary and most splendid impressions so that you must not impute to me that this letter has an even more fragmentary shape than the preceding ones. Because at every moment I succumb to the temptation to get up from the table and take a look out of the window."[5] What he saw from the window were no antiquities but the immaculately white houses of Phira, the island's capital, that cascade down the crater's precipice. Some 700 feet below, in the midst of a vast bay, he could make out Thera's eternal menace, three volcanic islands rising from the wine-colored sea.

In 1866, after a centuries-long pause, Thera's volcanoes resumed their activity, thereby inducing a great number of observers to study them at close range. While digging test pits, vulcanologists chanced upon prehistoric dwellings of a type that still prevails in the island. Yet none of these inquisitive men had an eye for them. Thera's

198. The dependence on a single building type does not necessarily produce monotony. These Theraen houses form a striking contrast to the official rubberstamp design of recent years. *(See figs. 203 and 204 on p. 243.)*

peerless contemporary vernacular eluded them completely. "The keynote of all their reports," I recounted elsewhere, "was *primitivity* (. . . no carriage that ever made the acquaintance of Thera's desperate cinder paths; no stove to mitigate the winter cold, not a tree to dispense coolness in summer . . .) but this was not a sign of backwardness due to the long Turkish rule. Rather one ought to speak of *perfection* as expressed in a marvellous unity of past and present, which imparts to the stranger a sensation of timelessness."[6] The savants seem to have overlooked the fig trees, olive trees, and the tree-high cactuses and, more particularly, Thera's vineyards. At least the wine from the grapes that grow in the ashen soil, famous since time immemorial, ought to have made up for the shortcomings, real and imagined. Morcover, the visitors could easily have found out that the island's troglodytic and semitroglodytic houses obviate stove and cooling fan.

To make the most of their soft volcanic rock, the Theraeans tunnel into it, Chinese-style, and lay out several rooms, one behind the other. While some houses are half–rock-bound, as if emerging from a geological chrysalis, their rooms are not in any way different from those of ordinary, free-standing houses. These latter consist of one or several oblong cells, each covered by a barrel vault. What roots them in one's memory is their stylistic chasteness. The islanders never employed such conventional vaulting techniques as the corbeled roof of their neighbors, the Mycenaens, or the standard Roman arch. Skipping thousands of years of evolution in the building field, they hit, right from the beginning, at a method for making a strikingly modern-looking concrete shell. Today as in the past, the materials of this roof are the local pumice stone and pozzolana, a volcanic ash that, mixed with lime, yields an exceedingly firm hydraulic cement. The walls of the houses are built of lava blocks.

199. In the rectangular universe according to Cosmas of Alexandria, four walls carry the vault of Heaven. The hump in the center is the earth, washed by rectilinear oceans. The two dark disks represent the summer and the winter sun.

Although the barrel vault mainly triumphed in ecclesiastical architecture, it has many profane antecedents. Its spread was and still is near-universal. Every imaginable material besides stone has been employed in its construction—papyrus, palm leaves, osiers, clay, metal, and canvas. It found application in tunnels, tombs, and strongrooms, in the house wagons and travel wagons of India and China, in our railroad cars. However, it was left to a sixteenth-century monk, Cosmas of Alexandria, to invest the barrel vault with a celestial aura.

A one-time merchant, explorer, and traveler, mapmaker and geographer-theologian, Cosmas is remembered as the author of *Topographia Christiana*, the most elaborate of antiscientific treatises on the universe. A conscientious objector to the theory that the earth is round, he equally abjured the idea of a spherical Heaven. To his mind the earth was covered by a vaulted firmament, more precisely, a barrel vault. In order to reconcile his doctrine with terrestrial topography he postulated for the earth a rectangular shape. His graphical representation of earth and Heaven resembles, alas, nothing so much as a rural mailbox.

Unlike the barrel-vaulted Tunisian houses (fig. 200) which are deployed in military formation, those of Thera rise erratically. Here, the vault proliferates in its own intricate ways. No two houses are alike; the width of the roof is never the same. The vault sometimes curves steeply or it lies flat, or the long sides diverge from the parallel. Besides, many a roof, or at least part of it, is banked, forming a level terrace to better catch the rainwater.

200. Uncompromisingly utilitarian storage houses at Medenine, Tunisia.

201. The very sight of a hilltown fills the asphalt-bound urbanite with dismay; negotiating hundreds of steps in the course of a day often is beyond his mental and physical capacity. These luminous houses climb one of the twin hills of Siros, the capital of the homonymous island. Cyclades, Greece.

The single most engaging trait of these houses is what might be called their incandescence. In the towns and villages that border the Mediterranean, whitewashing is performed with the same gusto that Orientals bring to their ablutions. It is far more than an architectural cosmetic, although it does serve to cover up imperfections. The houses acquire their resplendent whiteness the same way lacquer acquires its hardness—by the application of innumerable layers—with the difference that no coat of whitewash is the last. Not only the walls, inside and out, are painted over but so are roofs, stairs, and doorstep, and sometimes the floor, indeed, the street in front of the house. On the Cyclades, peoples' faith in the godliness of cleanliness is perpetually upheld and demonstrated. On many of the islands, writes E. Bradford, "the pavements are white-washed twice daily. It is difficult to describe this atmosphere of cleanliness except to say that when you step ashore, you find yourself looking at the soles of your shoes before setting foot in the main street."[7] In some parts of Andalusia custom demands that the house front be touched up daily. After every meal the housewife goes with the paintbrush over door and window frames to let the flies know that they are unwelcome. Thera's towns, spectacularly set off against the volcanic earth, represent snow-white vernacular at its most radiant. In 1956, a violent earthquake nearly put an end to them.

This is not the place to enlarge upon Thera's convulsed vulcanological history, engrossing though it is. Besides, the island ranks high among the contenders for being a vestige of the legendary Atlantis, having been visited by what was presumably the greatest natural disaster since man walks the earth. A mishap of a different order hit the island in the aftermath of the latest quake—the inrush of a wave, not of water as happens after seaquakes, but of architects. With the towns largely reduced to ruins, the Greek government programmed ten groups of houses and one brand-new village to provide for the homeless. The Housing Department of the Ministry of Public Works was faced with the delicate problem of devising an up-to-date economical building type without doing violence to the local architecture.

202. An Andalusian housewife, smocked, trousered, and gloved, at the never-ending task of touching up her house's walls with whitewash.

The architects charged with breathing new life into the island towns had, like all architects, been trained to design rather than to build. Moreover, on Thera they had to compete with a local building tradition that never knew drafting board and ruler. It is of course the architects' business to make projects, and that is what they did for Thera. Still, to employ the architectural clichés of the day would have been inopportune. As a compromise and an obeisance to the island's distinguished heritage they decided to retain the local vernacular.

Nevertheless, in the course of planning, designing, and decision making, the traditional houses' languid lines were resolutely straightened out, their wayward curves stiffened into half-circles, the dimensions of rooms, doors, and windows standardized, and the foundations laid out perfectly parallel in homage to the Right Angle. That way, money was saved and work speeded up, but something precious got lost. With identical houses, ranged in rows, all individuality died in the good cause of equality. "Much of the housing appears to be laid out borough-engineer style," commented the *Architectural Review:* "the houses, while individually vernacular in appearance are grouped in a manner that is antivernacular visually, and may also prove to be so socially, as well."[8]

One would have expected the homeless families to have been grateful for a roof over their head—any roof, regardless of curvature and overall dimensions. Having just escaped maiming or death, it would have seemed presumptuous on their part to find fault with their shelters' appearance, least of all, to be sticklers for archaic forms. Yet whatever the validity of the Theraens' reservations, the streamlined version of the local vernacular was no success. Five years after the quake, even before the building program had reached completion, some of the new houses had already been deserted.[9] The people had moved to authentic island houses, built by private enterprise.

This frank aversion to living in uniform houses redounds to the islanders' credit. Their fastidiousness, nurtured by a magnificent environment, found its happiest expression in their old towns and villages where the physiognomy of a man's house is as unexampled as that of his face, where the vernacular holds its own as the antithesis of all planned architecture. "In the Greek village they didn't have to paint every house a different color," noted Moshe Safdie, the propounder of a contemporary vernacular; "people recognized their own houses because difference was the result of complex forms."[10]

At Thera the complex forms are caused to a great extent by an agitated topography. The houses are connected by stairs that spout like mountain brooks. The dancing rhythms of the steps, with their startling pauses, the landings, are rarely registered by the plantar nerves of those who all their life have been conditioned to walk like mechanical toys and thus lost the sprightliness that nature bestows on man and beast alike. Whereas in our buildings the stairs are marching up and down with metronomical precision, the unstandardized steps of light-footed people have their *ritardandi* and *rubati;* ramps are tangible *glissandi.* The varying height, width, and depth of the steps are the despair of the city dweller. To scale but a short flight, he needs the crutches of railings and bannisters. In fact, they are written into our building regulations.

203, 204. The national government's streamlined version of the island's vernacular was cold-shouldered by the earthquake victims. Which goes to show that country people have a much stronger attachment to their architecture than urbanites suspect.

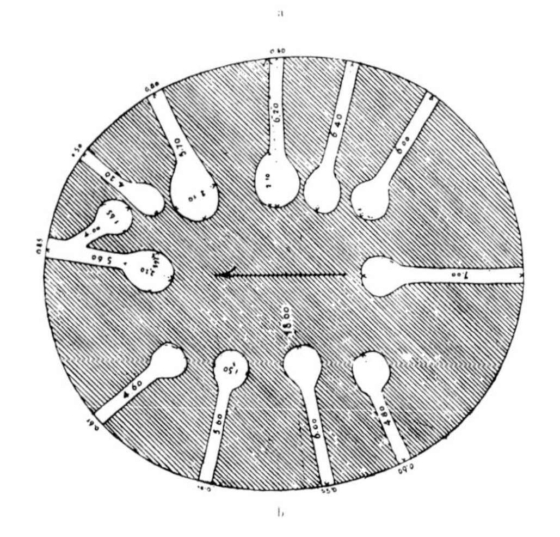

205. *Plan of the many-chambered neolithic Sese Grande on the island of Pantelleria. The axes of the oval are 58 and 66 feet long. (After Monumenti Lincei)*

206. *Pantelleria's dammuso is a prime example of agglutinative architecture; it grows in clusters like chanterelles. The near-identical units connect directly, yet the absence of corridors does not prejudice privacy.*

A runner-up in the category of massive architectural folklore is Pantelleria, an island about midway between Sicily and the Tunisian coast. Like Thera, it is of volcanic origin; like Thera, it has a population of about ten thousand, distributed over 35 square miles. Violent winds bring punishment to humans, beasts, and trees, whose only defense are thick walls of obsidian blocks, constructed *a secco,* that is, without mortar. Like every respectable Mediterranean island, Pantelleria, too, has its neolithic village and neolithic towers, which here are called *sesi*—squat round and oval constructions, somewhat similar to Sardinia's nuraghi.

Of more immediate interest to the casual visitor are the native peasant houses, called *dammusi,* mint specimens of architecture without architects. The dammuso is Pantelleria's patrimony. Although it still carries the day over the philistine's villa, its future does not look bright. Pantelleria is a victim of twentieth-century technology. Because the island's chief product, wine, cannot compete with the mainland's bogus wine of chemical manufacture, one-fifth of the vineyards lie fallow, the land being gobbled up by the tourist industry. By the mid-1970s, fifteen hundred houses, some quite extensive and in good repair, had been abandoned by their inhabitants who had left to earn their livelihood in the cities.

The empty dammusi are too unprofitable to be marked for demolition. Luckily, they are among those architectural strays that caught the imagination of urbanites prospecting for congenial rural *Lebensraum,* thus reversing the trend of the Panteschi, the island's inhabitants. New residents are recruited by word of mouth. These days a band of intrepid young architects, less complacent, less venal than their elders, have taken it upon themselves to rehabilitate the indigenous houses.

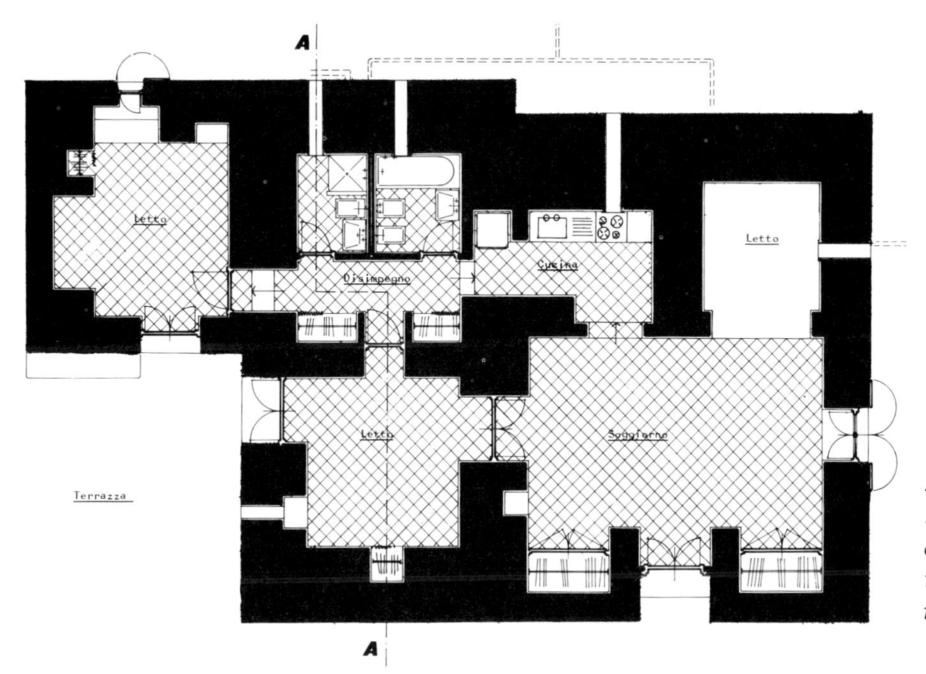

207. Plan of an updated dammuso, a house type native to Pantelleria. Closets and shelves are built into the walls' thickness. So is the stone platform letto, *or bed, in the upper right corner.*

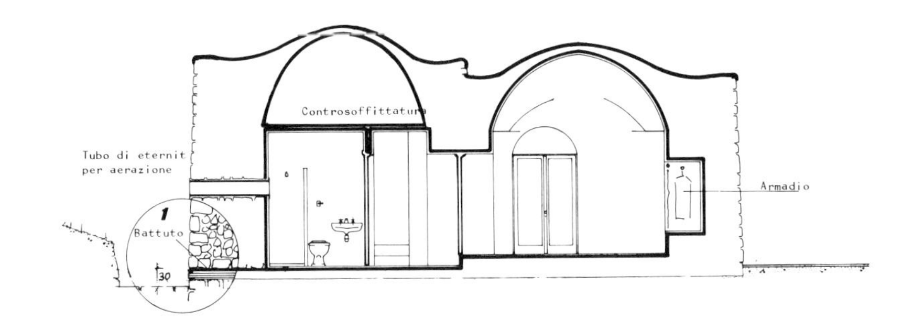

208. Section A—A of the dammuso in figure 207 reveals graceful roof construction, up-to-date fittings, a recessed clothes closet, and a peephole for a window.

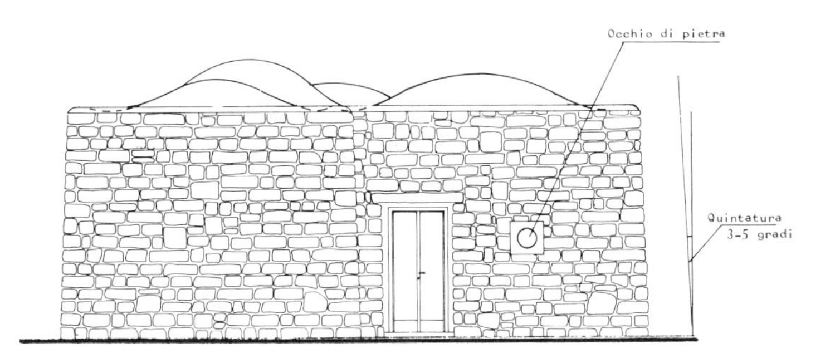

209. The only openings in the walls of volcanic stone are doors and occhi di pietra, *round slits for admitting daylight of African intensity. (Drawings courtesy Gabriella Giuntoli)*

If Pantelleria merits a star in a compendium of vernacular architecture, it is for yet another reason. The winds, scourging the island as pitilessly as Marseille's Mistral, have given occasion to a unique primitive building type, popularly referred to as *giardino*, meaning garden. Here, however, the word designates a kind of miniature fortress, scattered about the vineyards. The giardino is a massive ring wall, about 15 feet high and, like a dammuso, built of mortarless stone masonry. Usually it describes a circle, but there also are octagonal and square ones. It has but one entrance, firmly barred by a wooden door. Some day antiquarians will quarrel over this native peculiarity, and identify it as anything from an offshoot of the island's prehistoric sesi to a mid-twentieth-century machine gun nest. (Pantelleria was heavily fortified in World War II when it figured as a stepping stone in the Allies' invasion of Italy.) Actually, the giardino represents an unheard-of extravaganza, a single-tree orchard, whose sole purpose is to protect a lone lemon tree—a species far less robust than an orange tree—from the furious winds. The tree is neither a cult object nor a horticulturist's showpiece. It produces a thousand or more lemons a year, indispensable for an Italian household. In the giardino rural architecture achieves the simplicity of a heraldic device. It embodies the archetype of the paradise (a Persian word meaning circular enclosure), complete with the tree of sour knowledge. But basically, it is an arboricultural solarium, with room to spare for a latter-day Adam and Eve.

210. *Pantelleria's architectural oddities are circular stone-walled enclosures, each of which acts as a windscreen for a lone lemon tree.*

211. The all-white, bosomy, benippled trullo with a square interior shows a late stage in the slow changes of this building type. Originally, the conical roof rose from a round structure.

Similar rustic bastions are encountered in southern latitudes. For instance, on Hierro, the westernmost of the Canary Islands, fig trees are surrounded by high walls, not against stormy winds but against the predatory goats. Each tree represents a patrimony. Like the vines of Beaune and Nuit-Saint-Georges, one plant may be owned, branch for branch, by several families.

Isolation, maintained Darwin, is of great importance in the production of new species. He didn't have architecture in mind, but his pronouncement applies to it as well. Typical examples are the aforementioned prehistoric structures. So is the *trullo*, a stone house native to parts of Apulia, near the heel of the Italian boot. So far it has survived, if precariously, progress's kiss of death and the hazards of commercialism. It even endured its promotion to a national monument, an honor not lightly conferred these days, least of all to an oddity without any standing in the architectural hierarchy.

For a rural dwelling of hoary lineage the trullo is surprisingly appealing. Blazingly white from top to bottom or, when more conservatively appointed, with only its walls whitewashed and its stone roof left *au naturel*, it has none of the grimness one expects from an archaic habitation. On the contrary; the most fitting adjectives for it would be pretty or idyllic, none of which have currency in classical or modern architecture. Its luminous presence is underscored by bucolic surroundings, a patchwork of vineyards, cherry orchards, olive and almond groves, presenting a dazzling spectacle at blossom time.

This is of course a roseate account; people hailing from Nordic countries usually see the south through dark glasses, actually and figuratively. An English archaeologist, reporting in a professional journal on the Apulian dwellings, found the scenery "unattractive," the country "bleak and desolate," with "stone in embarrassing profusion." Alberobello, the trulliesque town, struck him as "unreal and wildly fantastic."[11] Apparently, the charms of southern folk architecture manifest themselves only to warm-blooded creatures.

212. Seen from above, the gray roofs of trulli call to mind herds of pachyderms. The roofs are noteworthy for their pliant, tentlike forms, obtained with rough stone. Alberobello, Apulia.

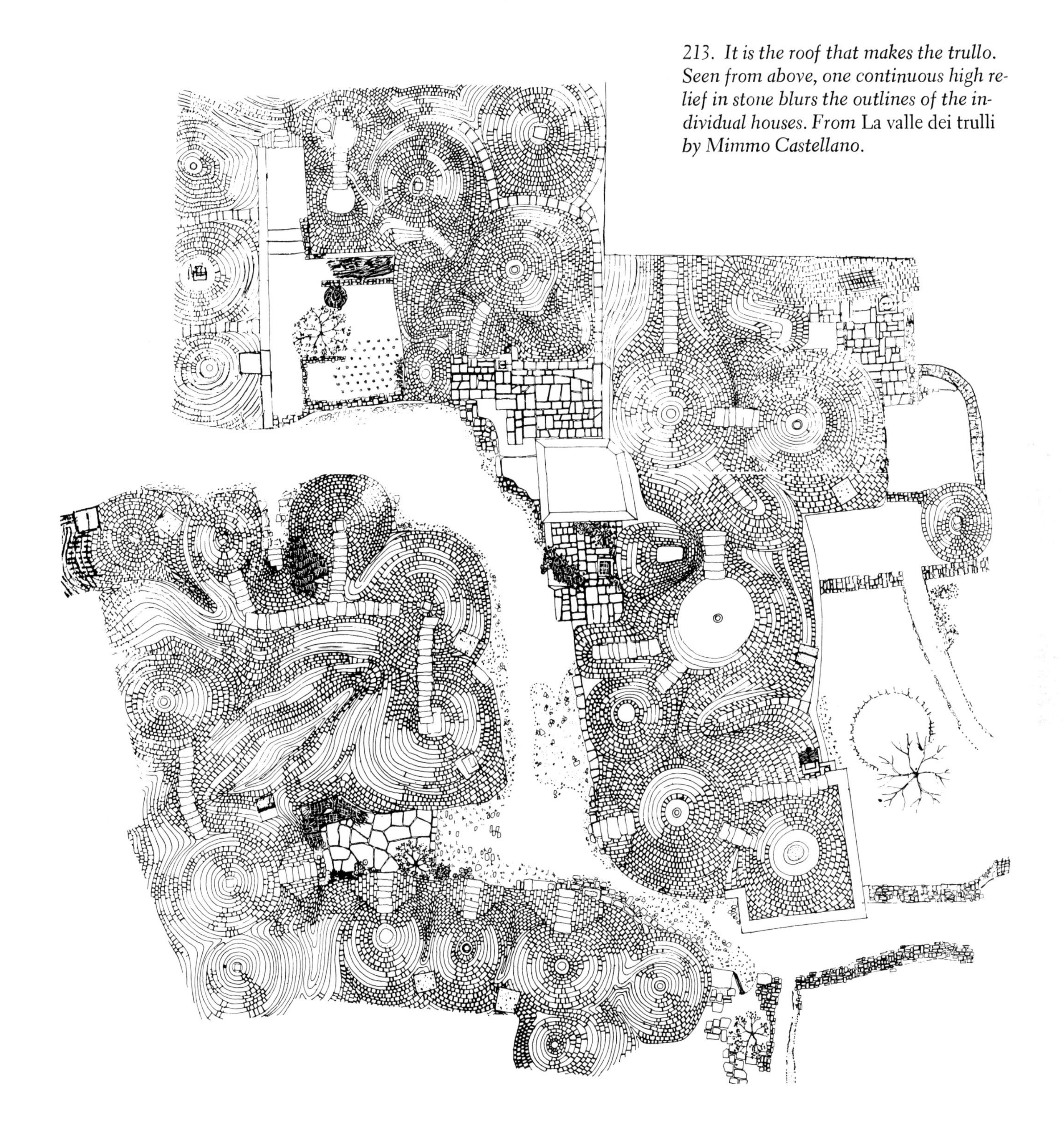

213. *It is the roof that makes the trullo. Seen from above, one continuous high relief in stone blurs the outlines of the individual houses. From* La valle dei trulli *by Mimmo Castellano.*

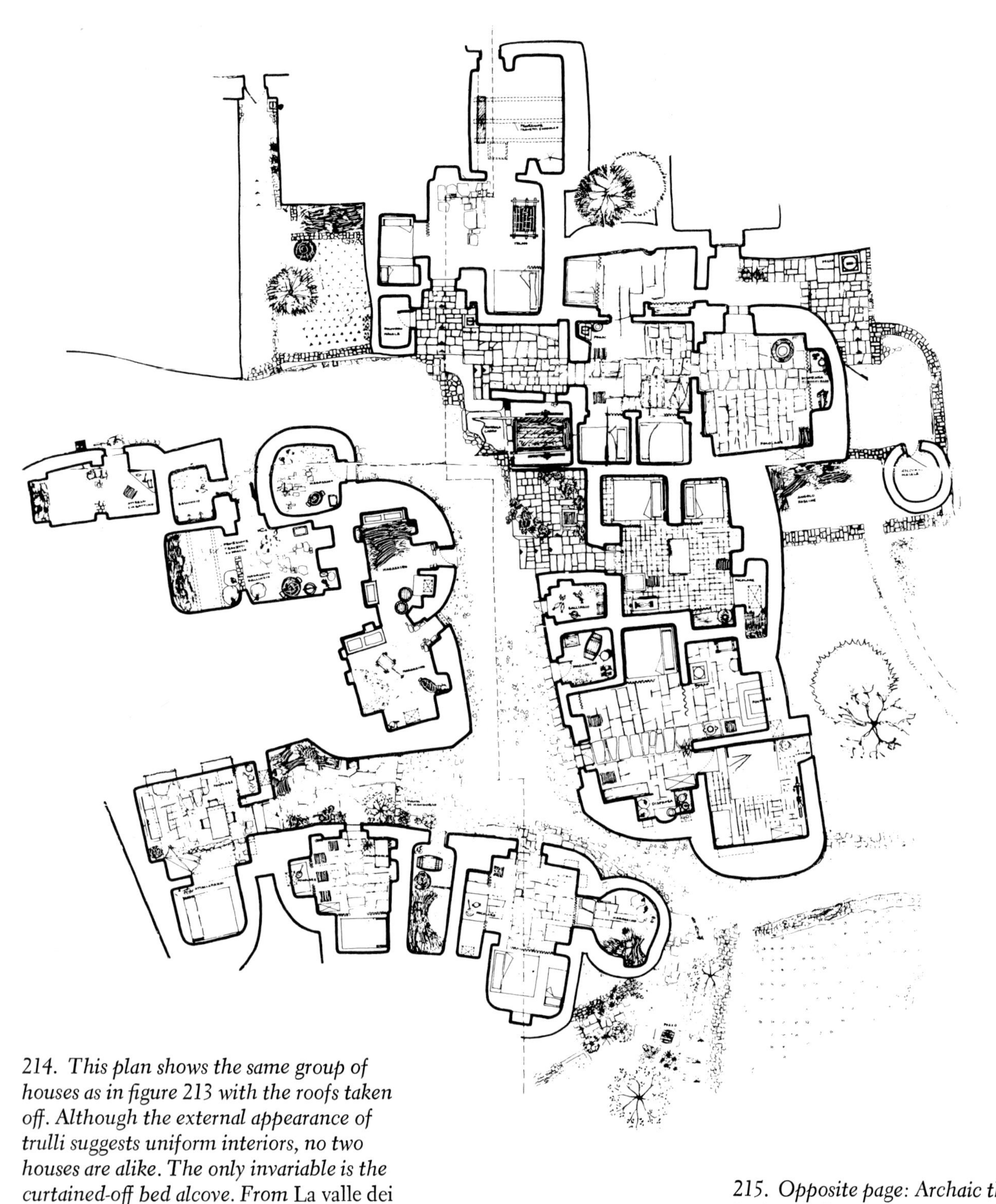

214. This plan shows the same group of houses as in figure 213 with the roofs taken off. Although the external appearance of trulli suggests uniform interiors, no two houses are alike. The only invariable is the curtained-off bed alcove. From La valle dei trulli *by Mimmo Castellano.*

215. Opposite page: Archaic though it may look, a trullo is in some respects superior to the conventional house. Its stone walls and stone roof make it incombustible. It keeps warm in winter, cool in summer.

Strictly speaking, the trullo (from *truddu,* Greek for cupola) is not a building type but a manner of construction. (Trulli of sorts are found on the west coast of Sweden in the district of Halland, though the paternity, if any, has never been established.) The trullo is built on a square or rectangular plan but occasionally veers off into a would-be circle. Like most vernacular architecture of a becoming awkwardness, it is innocent of a plumb line. The walls of the older buildings are often as many as 10 feet thick, although in conformity with the progressive shrinking of domestic architecture's marrow, people are now content to put up with lesser dimensions. Paradoxically, the trullo's roof is not pyramidal—which would be the logical form for covering a square—but a cone, or rather a liberally interpreted cone shape that turns at the slightest obstacle into a nonrecurring, nondescript form. What look like shingles are circular layers of stone which form a retrocedent wall rather than a roof. The inside is a parabolic vault of larger stones, the spaces in between filled with rubble. As

216. Trullo section: (1) storage; (2) all-purpose room with bed alcove; (3) cistern. From La valle dei trulli *by Mimmo Castellano.*

befits a structure that traces its ancestral line to the beginnings of architecture, no mortar is used.

Sometimes the roof comes truncated, or rises in several steps, resulting in profiles which confer on it a severely millenarian air. Occasionally, an outdoor flight of narrow stairs leads to the top, adding another incongruous touch. The roof plan of a cluster of trulli (fig. 213) conveys with its swirling lines the gracefulness of an architecture that for centuries remained unappreciated, not to say invisible. Such a discerning man as the philosopher George Berkeley, the first foreigner to venture into Apulia (in 1717), took no notice of it; he saw in the trulli but "round and pyramidal heaps of stones."[12] Their discovery occurred fully two hundred years later.

Trulli have a tendency to sprawl. At the arrival of a new child a new unit is added, just as in African latitudes a new hut is built for each additional wife. The denizens of the trulli are emphatic about their houses' merits. Thanks to a dry construction method, the houses are healthy. They also are economical because nature stockpiled the building material right on the site. The thick walls are pocketed with deep alcoves into which are tightly fitted hearth, sideboards, and beds. In piquant contrast with this penury of space stands the monumental ceiling, the pseudocupola of voluptuous contours that was

217. The naked wintry landscape near Martina Franca, Apulia, emphasizes the toylike character of the trulli: near-perfect cubical houses with white cones for roofs.

once the last word at Orchomenos and Mycenae. Trulli are truly cousins of *tholoi*, the Mycenaen tombs. Their affinity with the archaic constructions on the Greek mainland could be easily construed, for Taranto, the ancient Taras, in whose neighborhood stand the mightiest of trulli, once was one of the richest and most powerful of greater Greece's towns.

Today, real estate, renting, and travel agents cast a greedy eye on the quaint little houses, thus for the first time in history endangering the species. Progress came to Apulia in the postwar fifties and left its mark. A number of houses suffered improvements such as tightening their contours, painting them in the colors of the rainbow, and availing them the decorative attributes of the suburban bungalow. Nevertheless, a good many of the architectural fossils that escaped the facelifting had their life expectancy prolonged by becoming collectors' items. More recently, an enterprising man built at Alberobello a motel in bogus vernacular—a modern caravanserai in neo-neolithic style.

218. Besides artifacts shown in folklore museums there exists a nonportable rural art practiced by humble workmen. The stone masons who composed the granite and stucco mural above the stable of this Portuguese house were the unsung Arps and Hélions of generations past.

The vernacular obliquely appraised

For historical and geographical reasons English-speaking nations have a deplorable mental picture of rural architecture and peasantry. To them both are abjectly alien. *Peasant* connotes a serf, a low fellow, a rascal; it is a term of abuse. The word *peasant*, the *Oxford English Dictionary* states, is in use, "properly only in foreign countries." It follows that peasant architecture is a product pertaining to foreign countries only.

The millions of peasants who immigrated to the United States stopped being peasants the moment they stepped ashore, for they never had an opportunity to make use of their know-how. The brio they brought to their work in the home country got lost in transit. Moored among an uncongenial humanity, reviled and humiliated, they spent the rest of their years eking out a precarious existence in the cities. "They cut themselves off from their cultural roots, because that was how you became an American," said the head of the Task Force on Urban Problems.[1] The blessed earth was left to be cultivated by an industrial worker, the farmer.

The country would represent a different face today had it been entrusted to competent husbandmen. Instead it was surrendered to the pioneers. These men who found no great cities, no fortresses, no works of art to destroy for the glory of God, had to be satisfied with exterminating the fauna and denuding the forests. They had no love for the land and the things that grew on it. The soil held no mystical attraction for them. Trees found no favor in their eyes. Each year, loggers caused fires to as many as 25 million acres; "the common assumption," wrote former Secretary of Interior Stewart Udall, "was that trees, like Indians, were an obstacle to settlement."[2] Besides, the people who tilled the land did not know the first thing about farming. "Not until the beginning of the twentieth century," S. F. Markham pointed out, "did agriculture generally in the United States regard soil conservation as essential to good farming, and even then for thirty or forty years few of the best methods were put into practice."[3] Such a basic measure as contour plowing was introduced only a little more than a generation ago, yet it had been in force in China for five thousand years.

Paradoxically, it was North America's very virginity that discouraged tender feelings toward Nature. ("A hideous and desolate country," it was called by William Bradford, a Mayflower passenger and colonial governor.) For what draws man emotionally to the outdoors is less Nature's grandeur than the man-made landscape. The majesty of a glacier or a deep forest affects the most stolid person but only in the way a spectacular sunset or a thunderstorm do—phenomena out of scale and out of reach for an intimate experience. What touches the heart is the mark left by the man who cultivates the land, and cultivates it wisely; who builds intelligently; who shapes his surroundings with a profound sense of affection rather than in the pursuit of profit. What he treasures are the many landmarks, from the elementary to the complex, that add up to the physiognomy of the countryside: the fields and groves, and the walls that separate them; the mile-long lanes of shade trees by the wayside or the blackberry bushes shielding a sunken road; a footpath over the mountains that has served wanderers for a thousand years. Antiquated and more than faintly disreputable as a footpath seems to the man who depends on his car for locomotion, it still has its uses. A stroll—and we need not be concerned with either the mechanics or the therapy of walking—affords endless opportunities for taking stock of the world around us. Walking is often the only way to see what a country and its people are like. Chances are that one who never walks of his free will remains an ignorant person all his life.

By no means do the scenic features referred to have to be quaint to appeal to our hooded senses; the masts and wires of a high-tension line crossing a valley high up in the Alps agitate the intellect and prove to be as stimulating and aesthetically pleasing as a hallowed castle ruin. It is the human touch that counts, the happy touch, which is nowhere more perceptible than in what we please to call underdeveloped countries, where a tree is still a tree and not just lumber, where a river is not primarily a sewer; in short, where culture has left its mark.

219. *Opposite page: In industrial countries a wood, a grove, or an orchard tightly hugging a village or a town count among life's unattainable luxuries. This palm grove acts as the lungs of the town of Qatif, Al Hasa, Saudi Arabia. (Courtesy Exxon Corp.)*

220. The distinction between village and small town is largely legal, fiscal, or whimsical. The Portuguese Castelo de Vide is decidedly municipal according to Old World reckoning, yet hopelessly peasanty to the mind of people who expect to drive their car to their doorstep. Essentially, Castelo de Vide is one of those unplanned communities of undesigned houses which achieve harmoniousness despite diversity or what might seem chaos to the architectural disciplinarian.

City people gradually lost touch with Nature, so much so that eventually they came to regard it as their enemy. Yet once upon a time man stood in awe of the visible heavenly bodies rather than, as today, of a hypothetical, unperceived Heaven. He idolized the sun, the moon, and the stars without letting himself be distracted from the wonders of the earth. Man's earliest gods were personified natural forces. While we see in his reverence for Nature but superstition, pagan cults nevertheless persist among the peasantry. On the whole, however, the pantheon of pagan deities has long given way to battalions of plaster saints, the sylvan glades have been turned into picnic grounds. Hence the inability of the man in the street to perceive, let alone appreciate, rurality and, with it, rural architecture. Age, nationality, education or the lack of it; the meteorological and mental climate to which he has been exposed during childhood; professional ossification contracted in later years—all contribute to keeping distance from the products "pertaining to foreign countries."

Village culture never took roots in America north of the Mexican border. Schools of architecture that for centuries taught the pirating of ideas and forms three times removed never deigned to notice the rural scene. "When I went to school," reminisced Louis Kahn about his own pilferings, "we had a reference library divided into various architectural periods: Egyptian, Greek, Roman, Gothic, and so on, and this was my realm of architecture. If I had a cemetery to design, nothing could be better than a walk to the Egyptian era where I could find what I needed. That was just the kind of life I led, and it was most delightful; I looked through the books and saw wonderful examples I could follow. Now when I got through school, I walked around the realm and came to a little village, and this village was very unfamiliar. There was nothing here that I had seen before. But through this unfamiliarity—from this unfamiliar thing—I realized what architecture was. . . ."[4]

Alas, Kahn's cryptic account of his postgraduate discovery leaves us in the dark about the particulars. The soft-focus image of the "little village" remains indistinct, his walk around the realm has a sleepwalking quality. (He was unwilling to elaborate his parable.) Surely, to most of us, a village, real or imagined, is more exotic than an Egyptian temple. Any architectural monument, stone-dead though it may be, if properly classified, commands respect. Not so the village; it hardly arouses curiosity. And at the back of our mind there lingers an ever so faint distaste for the villager's communal institutions and goings-on.

Tribal spirit, so foreign to solitary urban man, is the peasant's lifeline; a man by himself is a martyr, says a Serbian adage. In former times common ownership of land, house, and tools cemented close

221. Opposite page: This rural community in central Japan has preserved its integrity; Main Street with its movie house, supermarket, and empty lot is still in the future.

bonds between villagers. So did life's misfortunes, war, conflagrations, and death. Even in our days, village culture draws its strength from people's willingness to perform community duties. The awesome beauty of Japanese mountain villages is due in large part to cooperative work. Neighbors help each other not only with bringing in the crops but with building and repairing houses. (After the Pacific War there was an abrupt decline of crafts and skills; masons, roofers, and plasterers now have to be hired for work which formerly had been performed by the men of the community.)

Far from Japan, in Liberia, customs were much the same. If a man of the Kpelle needed a house, he asked his fellowmen to help him build it. No wages were paid or expected. A banquet at which palm wine flowed freely took care of one's indebtedness.[5] Hired labor—the white man's invention—was unknown to them. So it was half a century ago when professional builders were neither available nor

222. Opposite page: "The mosque gets a new makeup." Communal teamwork here has all the earnestness of child's play. The wooden poles serve masons as scaffolding when resurfacing the structures after rains. From Lumière d'Afrique *by Raphaël Mischkind.*

223. A communal life style used to have its compensations; the sociability engendered at an Italian town's monumental washtub (fed by a wellhead) compares favorably to the dreary atmosphere of the laundromat. This old-time laundry was located at Marino in the Alban Mountains near Rome.

needed. "A more animated scene than the thatching of a house in Fiji cannot be conceived," wrote one Captain Wilkes, member of a U.S. Exploring Expedition. "When a sufficient quantity of material had been collected round the house, the roof of which has been previously covered with a network of reeds, from 40 to 300 men and boys assemble, each being satisfied that he is expected to do some work, and each determined to be very noisy in doing it—when all are getting warm, the calls for grass, rocks, and lashings, and the answers, all coming from 2 or 300 excited voices of all keys, intermixed with the stamping down the thatch, and shrill cries of exultation from every quarter, make a miniature Babel, in which the Fijan—notoriously proficient in nearly every variety of halloo, whoop and yell—fairly outdoes himself."[6]

Such rural cooperation goes back, it has been suggested, to the time before the invention of bricks. As long as man wove his hut from branches, he was, much like a bird, self-sufficient. When his activities extended, he became dependent on the help of others. The perennial charm of peasant architecture lies in not having been homogenized; unlike formal architecture, it never degenerated into an Esperanto. Country people rarely go international. Hence a traveler is not likely to mistake an Andalusian village for a Swiss one or a Japanese hamlet for a Mexican one. Yet modern travel—to use a current euphemism for industrial tourism—contributes little to bridging the spiritual gap between town and country. Usually, what people know about, say, Sicilian peasants is still largely based on impressions gained from a performance of *Cavalleria Rusticana*. And in times before the ruin and rise of Hiroshima, their ideas of domestic Japanese architecture were mostly derived from the stage sets for *Madame Butterfly*. Today, the voluntary migrations of hundreds of millions of vacationers does indeed put them at arm's length of unfamiliar architecture. Increasingly, caves, castles, and ruins, generously drenched in *son et lumière*, are entering mass culture. Supernumeraries costumed as "folk" execute carmagnoles, tarantellas, and csardases against the background of impeccably groomed village squares. Or for a troglodytic experience, the tourist is herded into an Andalusian cave, tenanted by gypsies who make a living by nightclubbing folk music into acoustical pulp.

However, not every hamlet has been made over into a tourist trap, or so one likes to think. There are stretches of countryside whose climate is anathema to the urbanite; where roads are still in the future and the population is congenitally xenophobic. It is, as a rule, in these forgotten parts that one finds indigenous architecture in full bloom. Yet the farther one lives from places where folkways have withstood the inroads of misapplied technology, the harder it is to see

them for what they are. This was as true yesterday as it is today. To wit, John Murray's 1845 *Handbook for Travellers in Spain and Readers at Home* calls La Alberca with alliterative malice "a dark, dingy, dirty hamlet with prisonlike houses."[7] By contrast, the literate *Guide Michelin* praises the town's *cachet ancien exceptionel*. As it happens, La Alberca has long been declared a national monument in its entirety.

But then, remoteness itself is distasteful; the very choice of a

224. The plaza mayor, *a Spanish town's main square, often doubles as bullring and outdoor stage. At La Alberca in the province of Salamanca a sacred play is performed in honor of the* Virgin.

mountainside for the site of a village is incomprehensible to highway-conditioned man. His emotional response is contempt or wrath. D. H. Lawrence, a paragon of sensitivity *and* a passionate wanderer, equated Europe's Alps with Gothic fantasies in the eighteenth-century manner. Contemplating some sturdy peasant houses on the slopes of the Furka Pass, he could only think of impermanence and death. What he saw, or thought to see, were "the squattings of outcast people." In his ravings he equated *Bauernadel*—peasant nobility—with outcasts, sedentariness with squatting. "It seemed impossible," he caviled, "that they should persist there."[8]

These people, he argued with himself, "live in the flux of death,"[9] an opinion not confirmed by Swiss statistics on public health. His early topographical horizon having been defined by slag heaps rather than alpine meadows, life in a coketown appeared to him more wholesome, more desirable, than a rarefied mountain atmosphere. Could it be that he suffered from homesickness when traveling in peasant country? Apparently, he did; his hankering for the sight of an industrial landscape comes through loud and clear. "At the bottom of the valley," continues his description of the Furka Pass scenery, "was a little town with a factory or a quarry, or a foundry, some place with a long, smoking chimney; which made me feel quite at home among the mountains."[10] Englishness is a formidable hurdle to appreciating Continental ways. Had Lawrence done his homework, he might have discovered that on the Continent life in peasant houses is anything but melodramatic. Moreover, the Swiss and their neighbors are far from being encapsuled in poverty; they are doing themselves quite well. If he had wanted, Lawrence might have found, right in his home country, dwellings that corresponded to his lugubrious daydreams.

225. The multifamily apartments on Lewis in the Outer Hebrides are no Stone Age relics but local vernacular, a crude version of Mediterranean beehive houses. The windowless buildings are hardly distinguishable from hillocks. (After Thomas)

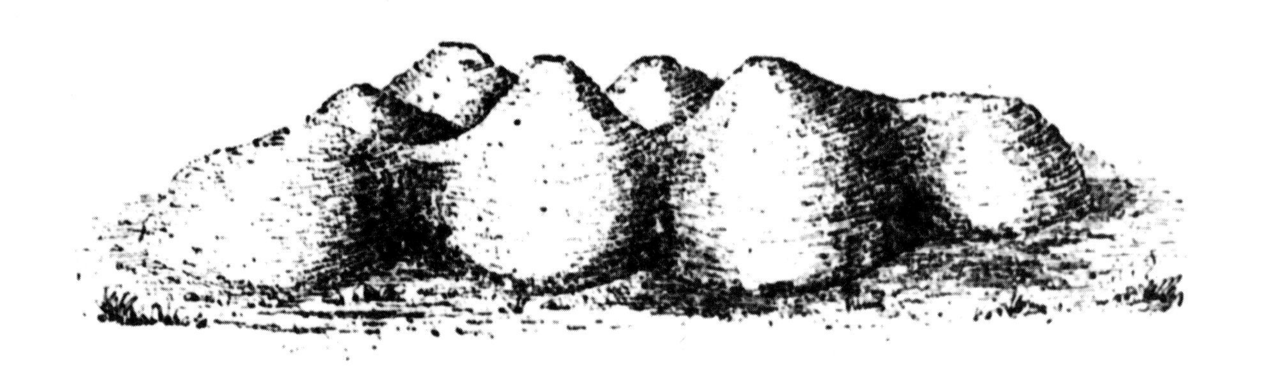

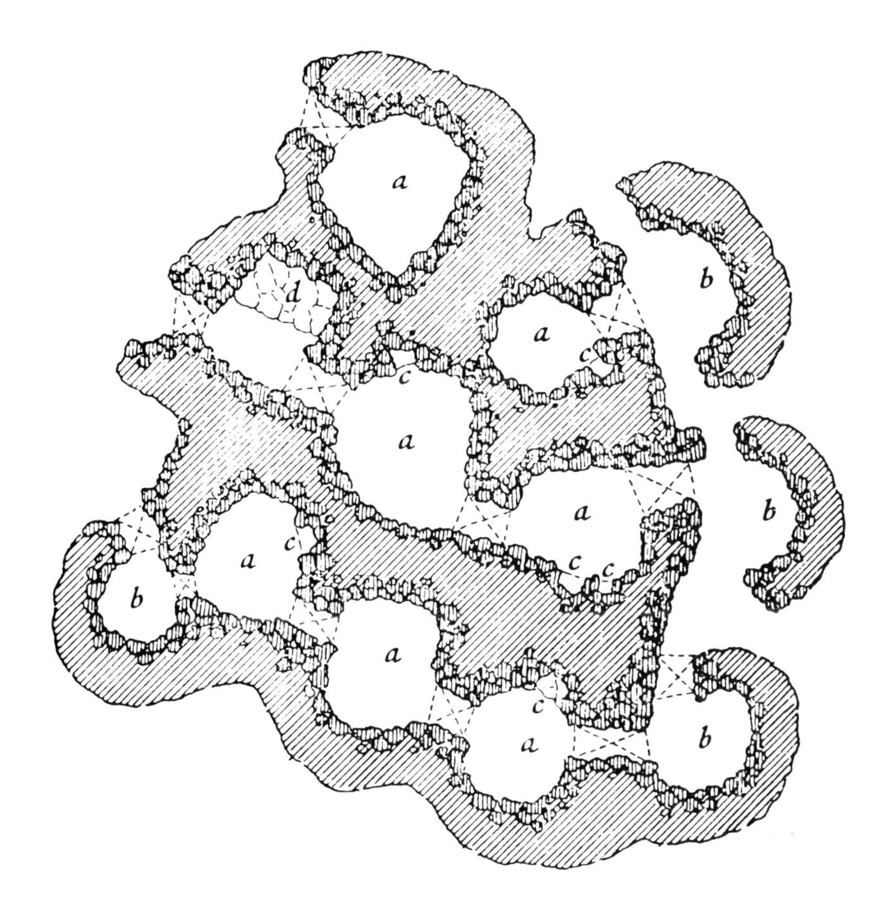

226. *Plan of the beehive houses in figure 225:* (a) *apartments,* (b) *porches,* (c) *milk cupboards,* (d) *a stone bed. From Mitchell's* The Present in the Past.

In Victorian times—and thereafter—living standards in the British Isles were spectacularly uneven, far more so than on the Continent. "We felt that we had been almost introduced to the stone age period."[11] one Arthur Mitchell, M.D., LL.D., Commissioner in Lunacy for Scotland, summed up (in 1880) his experiences in the island of Lewis-with-Harris. Some of the local domestic architecture was uncompromisingly simple, resembling "a Hottentot village rather than a hamlet in the British Isles."[12] The materials employed were stone and turf for which no working tools are needed. The houses' continuous roofscape (fig. 225) looked like excrescences of the ground rather than man-made structures, an impression reinforced by a public footpath leading over the roofs on which, Mitchell noted, children, sheep, fowls, and dogs ran about.

A similarly mixed company dwelt inside. Domestic animals provided warmth and an invigorating atmosphere, since "the portion of the dwelling is not cut off from that belonging to the brutes by the faintest pretense of a partition." Dung was removed only once a year. On that occasion, in districts where horses were used for transportation, the house, like a full piggy bank, was "often partially pulled down so that the horse and his panniers may enter to be loaded on the spot."[13]

The houses had no windows, "not even a hole." Nor did they have a chimney; the smoke escaped through the roof. From the sooty rafters hung a rope or a chain on which the cooking pot was suspended. Soot provided valuable manure, and its deposit within the house was therefore encouraged. A bench of sod or a plank supported by pieces of sod was all the furniture. Sleeping places were provided in the thickness of the walls. "I do not remember ever to have seen a table," noted the commissioner.[14]

He was not in the least dismayed by his discovery. Apparently, the primitive habitations did not adversely affect the inmates. The Lewis people, Mitchell pointed out with commendable patriotism, were physically, mentally, and morally well conditioned. In fact, he met with "more intelligence, culture, happiness, and virtue in those black hovels than in the houses of the Canongate and Cowgate of Edinburgh."[15] But he did not bother to adduce proofs of his findings.

The Harris houses were plushy compared to the shelter of a commune Mitchell encountered in the Wick Bay area. In a single cave he found two dozen people living the lives of happy savages who, he said, called to his mind Australia's Bushmen. Near the cave's mouth lay a man and a woman, "both absolutely naked," who apparently did not consider the appearance of an uninvited stranger enough cause for covering themselves. The others, in various stages of undress, were spread out on their lairs. For lack of furniture, stones served as tables and chairs. The cave itself was of no architectural interest, and Mitchell noted with regret the absence of cave art.

Neither Mitchell nor Lawrence were legitimate critics of architecture but their compatriot Ruskin was. In his essays on the "Poetry of Architecture," the ruminative Ruskin aired some of the national prejudices while leisurely discoursing on the merits of rural architecture, a far-out subject in Victorian times. Whereas he had honeyed words for the stones of great monuments regardless of place and origin, his milk of human kindness curdled before a mere *chaumière*. "There is," he wrote, "a general air of nonchalance about the French peasant's habitation, which is aided by a perfect want of everything like neatness, and rendered more conspicuous by some points about the buildings which have a look of neglected beauty, and obliterated ornament."[16] (If he had heard of the neglected beauty of Scottish hovels, he did not let on.) Although his utterances are not without irony, it is doubtful that his audiences were always aware of it. He made no bones about his countrymen's insularity; let us see, he wrote with tongue in cheek, "how far the Italian cottage agrees with our preconceived idea of what a cottage ought to be."[17] At one time or another, he seems to have fallen under the spell of the Italian peasant house. "All is clear, and warm, and sharp to the eye," he admitted; "the Italian cottage assumes with the simplicity, *l'air noble* of a higher order."[18]

Having conferred this consolation prize on the humblest of houses, he takes exception to its venerable age, as attested by its patina. To his mind, the moss and lichens that color the stones are inseparable from "the idea of decay." Like Lawrence, he saw most alien rural architecture drifting to perdition. At the same time, he resented the very slowness of this process. Detest as he did the peasant's instinct for conservation, he did not hesitate to equate endurance with stagnancy. "England," he pointed out, "is a country of perpetually increasing prosperity and active enterprise, but for that reason, nothing is allowed to remain till it gets old. Large old trees are cut down for timber; old houses are pulled down for the materials. . . . Everything is perpetually altered and renewed by the activity of invention and improvement."

227. The built-in obsolescence of our commercial architecture contrasts strongly with the virtual everlastingness of some rural constructions. In eastern Portugal stand houses built according to an archaic panel system. Sheets of granite, from 2 to 12 inches thick, are used in walls, roof, and pavements. The photographs show their application as wall panels mixed with ashlars. From Arquitectura popular em Portugal. (*Courtesy Sindicato Nacional dos Arquitectos, Lisbon*)

With what satisfaction must American readers have imbibed his words! What Ruskin was saying amounted to sanctioning a genteel kind of vandalism. The English cottage, Ruskin boasted, "is never suffered to get old, it is used as long as it is comfortable, and then taken down and rebuilt; for it was originally raised in a style incapable of resisting the ravages of time. . . ."[19] Here, then, we have in a nutshell the Anglo-American credo of built-in obsolescence and, implicitly, the scorn for the long-lived peasant house. Oldness, be it visited on stones or people, carries ignoble connotations in societies that worship youth and novelty.

These days, buildings have a shorter life span than the men who build them. The thought that a house might serve a family for several generations, and serve it well, has no currency. We accept premature architectural decrepitude as a matter of course. The culminating point in a building's life is reached the moment it is finished and its photographs taken for publication. In a way, this parallels the once popular belief that a person's days of youthful exuberance and venturesomeness end with marriage.

Poets and novelists seem more susceptible to the timeless vernacular than architects, if only because their judgment is less biased. André Gide, touring the Congo, was astonished that "the few rare travelers who have spoken of this country and of its villages have only thought fit to mention their 'strangeness.' " The Massa's hut, he declared, is not only strange; "it is *beautiful*" (his italics). It is so perfect, Gide noted, that it seems to have been arrived at mathemati-

228, 229. *Extremes of improvised shelter are exemplified by the alpine refuge in the Serra da Estrela, Portugal's loftiest mountain chain (opposite page), and the cardboard house erected by vacationers on Turkey's Mediterranean shore (left). Figure 228 from* Arquitectura popular em Portugal. *(Courtesy Sindicato Nacional dos Arquitectos, Lisbon)*

cally. Yet no forms, no scaffolding are used in its construction. "This hut," Gide wrote, "is made by hand like a vase; it is the work, not of a mason, but of a potter."[20]

Gide did not stop at admiring the hut's elegant shape but proceeded to inspect the interior. His description is worth quoting at length: "Inside the hut the coolness of the air seems delicious when one comes from the scorching outside. Above the door, like some huge keyhole, is a kind of columbarium shelf, where vases and household objects are arranged. The walls are smooth, polished, varnished. Opposite the entrance is a kind of high drum made of earth, very prettily decorated with geometric patterns in relief, painted white, red and black. These drums are rice bins. Their earthen lids are luted with clay, and are so smooth that they resemble the skin of a drum. Fishing tackle, cords, and tools hang from pegs; sometimes too a sheaf of assagais or a shield of plaited rush. Here, in the dim twilight of an Etruscan tomb, the family spend the hottest hours of the day; at night the cattle come in to join them—oxen, goats, and hens; each animal has its own allotted corner, and everything is in its proper place; everything is clean, exact, ordered."[21] Gide's words contrast pleasantly with those of tourists who decry the "filthiness" of all primitive architecture.

The gap between townspeople and peasants was never as pronounced as it is today. In former times, a taste for country living, and a curiosity about the earth's cultivation, had not been beneath a city gentleman but were thought of as highly becoming to him. Two thousand years ago, husbandry and rustic architecture were considered legitimate literary subjects. The books on agriculture by the statesman Marcus Portius Cato, and a man of letters, Marcus Terentius Varro, have not grown stale. (It was from reading Cato rather than today's survival scrapbooks that I learned to have my olives pressed immediately after gathering to prevent the oil from spoiling.) And Pliny's quasi-global outlook was not in the least diminished by his being concerned with the storage of corn.

Neither did Vitruvius, chief authority on classical architecture, deem it unworthy to expatiate on rustic buildings. Dictatorial when expounding on formal architecture ("In perfect buildings the different members must be in exact symmetrical relation to the whole general scheme"[22]), he takes on an almost human complexion when he gets around, in his *Sixth Book*, to the subject of such imperfect buildings as farmhouses. "From the point of health," his pastoral begins—and it is perhaps worth keeping in mind that his viewpoint rarely figures in pedigreed architecture where monuments are, as a rule, immoderately *unhealthy:* breakneck temple ruins, spine-chilling cathedrals, combustible theaters, draughty châteaux, and so forth; in short, hazards to health. From the point of health, then, Vitruvius advised, "let the kitchen be placed on the warmest side of the courtyard [as yet North Americans haven't discovered the courtyard], with stalls for the oxen adjoining, and their cribs facing the kitchen fire and the eastern quarter of the sky, for the reason that oxen facing the light and the fire do not get rough-coated. Even peasants wholly without knowledge of the quarters of the sky, believe that oxen ought to face only in the direction of the sunrise."[23] Which present-day Vitruvius has a knowledge of either oxen or the quarters of the sky! What is good for the ox is good for man. Early and primitive man knew all along that sleeping quarters ought to face east.

But perhaps nowhere in the world have gentlemen been as devoted to country living and its mainstays—horses, dogs, and the lower classes—as in England. Not only do Englishmen hold an enviable record as philanthropists and connoisseurs of nature, they have an eye for the beauty of (English) rural buildings. With local building traditions often sadly lacking, they have been known to apply themselves to the advancement of a desirable rural style with as much enthusiasm as their congenital reserve permits. In an album called *Views of Picturesque Cottages with Plans selected from a collection of drawings taken in different parts of England, and intended as hints for the improvement of village scenery*, published in 1805 by one William Atkinson, architect, are set forth, implicitly, some of the noble sentiments of the landed gentry: "The building of cottages for the laboring classes of society, and the keeping of them in good repair, are objects of the first national importance; as it is from the active exertions of the industrial laborers, that the other classes derive the greater part of those benefits which they enjoy."[24]

Why, indeed, should the lover of architecture concern himself only with the subtleties of capitals and fluted columns! Thatched cottages ought to have long been a national matter of consequence. Cottagers aquiver with sweet-home pride, if well taken care of, were known to repay kindness in kind. But the author of *Views* never descends into the maudlin. He shows a perceptiveness of social and aesthetic problems far above that of his fellow countrymen. "It is," he blueprints briskly, "a matter of high moment then, to the rich and opulent, that the habitations of their dependents should become an object of their concern; especially, as the necessary improvements might be made with very little trouble, and at small expense." Alas, although, as Atkinson pointed out, "the study of picturesque village scenery opens a boundless field of amusement,"[25] his attempts at selective breeding of cottages from which a modern stock may be derived never achieved results comparable to Continental rural architecture.

One man of the pen who did take up the axe to give shape and substance to his cottage reveries was Thoreau, and he actually came up with a passable rustic cabin whose construction cost, he proudly pointed out, came to no more than twenty-eight dollars and twelve-and-a-half cents. The most celebrated and least influential of nineteenth-century American structures, built more or less within shouting distance of the "luxurious and dissipated who set the fashions which the herd so diligently follows,"[26] it has long been reduced to dust, or rather humus. However, it lives on in *Walden*, the classical do-it-yourself treatise on domestic architecture.

In *Walden*, Thoreau expostulated on the mental discomfort endured in Western civilization through that self-imposed straight-jacket, our house. The lakeside cabin was his carpentered manifesto. His notions about what makes a dignified shelter were far more sophisticated than those of his contemporaries. Not given to raving about thatched roofs or herbaceous borders, he had instead a discerning eye for a house's inside. "Why," he asked among other things, "should not our furniture be as simple as the Arab's or the Indian's?"[27] The question was as impertinent as asking why a Boston dowager should not appear at the dinner table in her underwear. Not until a century later did the luxurious and dissipated opt for more spartan furnishings, by then made attractive for them by being exorbitantly expensive.

Unlike his fellow Americans, Thoreau preferred the primitive and austere to the opulent and garish. Against the "carloads of fashionable furniture," the potted palms and dowdy curtains, he held up the artlessness of a Bethlemitic stable. "I sometimes dream," he wrote, drifting still farther away from the domestic clichés of the day, "of a larger and more populous house standing in a golden age, of enduring materials, and without gingerbread work, which shall consist of only one room, a vast, rude, substantial primitive hall, without ceiling or plastering, with bare rafters and purlins supporting a sort of lower haven over one's head, . . . a cavernous house, wherein you must reach up a torch upon a pole to see the roof; where some may live in the fireplace, some in the recess of a window, and some on settles, some at one end of the hall, some at another, and some aloft on rafters with the spiders, if they choose; . . . where you can see all the treasures of the house at one view, and everything hangs upon its peg that a man should use; at once kitchen, pantry, parlor, chamber, storehouse and garret—a house whose inside is as open and manifest as a bird's nest."[28]

These were strange musings. A mid-nineteenth-century American, and a nonarchitect to boot, extolling the charms of a cavernous home, was news. In a world of narrow-chested houses with overcrowded

230. Opposite page: Thoreau might have found his dream come true in rural Japan—"cavernous" houses consisting of a single room as vast as the rigging loft of an opera house, unencumbered by furniture, its carpentry exposed. The illustrated old house is typical for the mountainous region of Takayama.

chambers of the most ungenerous proportions, Thoreau's cabin must have seemed to his readers downright peasanty, not to say barbarian. Whether the dreamer was conscious of this or not, it had its antecedents in the alpine loft, the Spanish *fonda*, the Japanese *gassho*, to name a few. In the year that *Walden* was published, Commodore Perry invited himself at gunpoint to Japan where he found the noblest mansions every bit as vast and bare as Thoreau's dream house.

Despite their contradictions, Thoreau's architectural fancies amount to much more than one man's vision of a contemporary vernacular, albeit of literary design and workmanship. They are a declaration of independence from the tyranny of drafting board architecture. Unfortunately, his example, and the ideas that led to it (he pleaded for a one-day work week, maintaining that the less man worked, the better for him), were too far ahead of his time to find a response in a nation with a built-in work ethic and an unquenchable thirst for mechanization.

And yet, if we think of a man as poor because his house and its appurtenances are not up to the advertised standards of the moment, what are we to make of affluent people's periodical escapes from their accustomed comfort? Are the pampered urbanite's sorties into the wilderness, his camping trips with their attendant hardships, merely recurrent masochistic fits? Is lodging in primitive mountain refuges, or sleeping under a draughty tent, due to atavistic impulses, unredressed by high living? Surely, discomfort *is* one of the symptoms of poor living, yet it does not necessarily constitute poverty. There exist boundless gradations of habitableness, from Diogenes' barrel to the *maisonette* extolled in fashionable periodicals. As it happens, we have no absolute criteria for measuring levels of penury in terms of domestic architecture. If we rate running hot water and electricity as minimum requirements for even the most humble home, Versailles and Windsor in their heyday would seem to have been just oversized hovels. Future generations may very well look back at our vaunted living standards as subhuman.

The future of vernacular architecture looks none too bright; the continuous extinction of native species was already observed at the beginning of the century. At that time a mild and somewhat erratic interest in what was commonly called folk architecture sprang up in northern Europe. It was caused by concern with the rapid loss of those traditions that for ages had imprinted an unmistakable identity on the land and its people. While the mountains and fjords stood unalterable, the form and substance of the house of man was undergoing a disturbing change. The distinctive buildings of old were increasingly giving way to bland, impersonal structures. Traditional building materials and building methods were becoming a thing of the past. So was that easy rapport of architecture with its surroundings. In districts with deep cultural roots and a rich heritage of folk art, people looked at the changes with a sinking heart.

Some practical men, however, motivated by much the same sentiment that makes them want to save a stray cat, sought out the proudest examples of rural architecture, pulled them down, and reassembled them in parklike environments which they designated open-air museums. Homesteads that for generations had been taken for granted were overnight promoted to museum pieces. The same attention that museum curators hitherto had bestowed on Belgian lace and Chinese lacquer they now extended to such humble artifacts as garden fences or the door of a hayloft. They also were able to prove that a lovingly wrought wicker screen had the edge on a barbed-wire fence. In the newly available aesthetic asylums, so fertile to the basic concept, was thus being celebrated the national genius of builders who never made architecture's *Who's Who*.

The oldest of these museums is Stockholm's Skansen, born in 1891, a miniature Sweden of genuine rural houses, stocked with characteristic household goods and costumed peasants. Between 1901 and 1912, Norway, Denmark, Holland, and Finland followed suit, each with their own folk art preserves. If European nations of more temperate zones lagged in collecting ruralia, it was not for want of a vernacular or indifference to its charms, but simply because the houses of peasants and fishermen had not yet become museum exhibits. In countries like Austria or Italy the need for salvaging specimens of native buildings did not arise because their rural architecture is still alive and well.

During the time between the two world wars the interest that had built up in vernacular architecture took a tumble. Modern architecture of every nuance completely eclipsed what had gone before it. The tradition-minded was denounced as reactionary or worse; handicrafts came in for ridicule, and ecology was a word found only in dictionaries. In recent years, however, architectural open-air museums have made a comeback. The most ambitious of these projects was undertaken in postwar Rumania.

After years of sociological spadework, a fund of buildings that documents the millennarian continuity of the Rumanian people (descendants of Roman war veterans who settled in Dacia after Trajan's campaign) were transported and reconstructed on a 20-acre lot in Herăstrău Park at Bucharest's outskirts. In the rustic setting of this Muzeu Etnografic in Aer Liber stand nearly three hundred authentic rural structures, arranged by geographic zones and set up on

231. A well-to-do Rumanian peasant's house from Salciua, Dejos, Alba County, built in 1815, now in Herăstrău's outdoor museum.

232. *Two eighteenth-century dugout dwellings from Rumania's Oltena province, on view in Herăstrău Park near Bucharest.*

scientific lines. They provided a direct link to archaeology, but chiefly they are of interest to an architecture-minded audience. Over eighteen thousand systematically collected objects are shown in and out of doors, making this village museum the most complete inventory of a nation's folk art.

It isn't just the diversity of the museum's colorful material that surprises visitors—homesteads dating from the early seventeenth century, dugouts of the plains, rustic mansions with pillared galleries and elaborate gates, a profusion of minor structures such as barns, stables, dovecotes, windmills, and water mills—what intrigues them most are the displays of preindustrial contraptions that once were essential to a farmhouse. At Herăstrău one is granted a peek into a world of fitments that have long disappeared: summer kitchens; outdoor ovens for baking bread or smoking food; presses for must and presses for oil from nuts and seeds; mills for grinding cereals; separators for milk and honey; stills and vats for fermenting plum brandy; fulling mills for processing woolen fabrics; looms; not to mention an array of agricultural tools unknown to the industrial farmer.

233. Barcelona's Pueblo Español is a medley of typical houses from all over Spain, tightly packed into 5 acres and presented like stage sets without consideration for scale or context.

But the museum amounts to far more than a catalog of archaic dwellings and the objects that went with them; it calls to mind forgotten ways of life. It teases one's senses with the ghosts of olfactory and gustatory sensations, vaguely remembered or never tasted—scents and flavors, familiar to past generations, that have all but vanished from the Western world along with the utensils for producing them. This tableau of unfeigned rusticity may seem too good to be true. Yet it reminds one that not all peasants were serfs; that many lived a better life than townspeople. Some houses with their opulent interior accountrements, their icons, thick carpets, and precious embroidered bolsters, radiate a downright Byzantine splendor.

Unlike village museums, Barcelona's Pueblo Español was intended, as the name implies, as a small town, albeit of a highly synthetic kind. This curious architectural hotchpotch was cooked up by several architects for the celebrated 1929 exhibition, where it shared honors with Mies van der Rohe's open-to-the-winds pavilion. Over the years the stone-and-plaster buildings have gracefully aged and taken on the patina of respectability. Along narrow streets and modest plazas, archetypical houses are strung up in overly picturesque array to give the rapid-transit visitor an instant orientation to the country's vernacular. None of the buildings is authentic. They are copies, mostly on a reduced scale, of existing landmarks that earned a star or two in guidebooks and touristic almanacs and figure prominently on picture postcards and travel posters. In some of the houses souvenirs and handicraft products are sold, but at least the make-believe rustics have been charitably omitted. Despite the aspiration after veracity, the trompe-l'oeil edifices are disturbingly incongruous, bringing to mind zoos where giraffes rub elbows with polar bears.

DE

The importance of trivia

Our domestic architecture has undergone much the same kind of purge as did that other vital commodity, food. To wit, in the past a premium was put on the genuineness and purity of all victuals. Eating was akin to a ritual, and the preparation of food was a welcome task. Food being considered a divine gift, prayer and blessing accompanied a meal. (No relation to that travesty, a White House Prayer Breakfast.) When the peasant, the customary provider of food, accidentally dropped a piece of bread, he picked it up and kissed it in a gesture of apology. Not only did he obtain forgiveness; the Lord granted him immunity from the bacteria lurking on the floor. The demands of a monstrously swollen world population put an end to food's sacramental status, and the art of eating gave way to feeding, with melancholy results. Moreover, largely depleted as the daily bread is of its nutrients, the chemical industry now offers a spectrum of pills for restoring to our body the very substances which the food industry removed.

A similar fate has befallen what we gullibly call progressive architecture. Advanced technology scorns Nature's gifts: The light of the sky, the radiant heat of the sun, the beneficial breeze are increasingly kept at bay. Although our glass houses suggest that we are confirmed sun worshippers, this is not the case. Glass merely happens to be a convenient, if expensive, building material that saves architects the trouble of bothering with the finer points of their art. The glassy skin, pulled tight over the skeleton, seals it against the outer world. To keep the

234. Opposite page: This abstract version of Laocoön and his two sons in the snakes' embrace symbolizes the mason's agonizing struggle with the clichés of classical architecture. The anguish and pathos expressed in this modest monument may afflict the beholder no less than the awesome realistic group of the ancient Greek sculptor.

inmate alive, he is provided with artificial respiration by way of air-conditioning and mechanical ventilation.

Once upon a time, sunlight and wind, much like food, were dispensed by divine agencies. People worshipped air gods and raised sanctuaries to them—to Boreas, the violent north wind, still known to and feared by the Italians as the Bora of the Adriatic; to Zephyrus, harbinger of spring; to the rain bringer Notos; and so forth. In return, spirits and sprites turned the wheels of wind and water mills and cleared the houses of foul air. Greek gods were part of Nature; there is no mention of the supernatural in Homer.

Whether one puts one's trust in Olympian powers or civil engineering, the four seasons are still around, and it pays to heed them. "A house should be built with the summer in view," wrote the fourteenth-century philosopher Yoshida Kenko; "in winter one can live anywhere but in summer a poor dwelling is unbearable." This much-quoted precept is not just a *bon mot* but a nugget of racial wisdom. No other nation equals the Japanese in coping with the summer heat. To air a house they simply remove its outer walls. Such a radical measure is inconceivable to us. It strikes us as an enormity, like tearing out an insect's wings. However, the operation is painless since the roof rests on pillars, not on walls. When the sliding wall panels have been shunted to a siding, the house seems to have lost its gravity. The fragile looking structure that emerges greatly resembles

235. The ingeniousness of the Japanese solution for ventilating the traditional house is unsurpassed anywhere. By removing some of the outer walls it becomes an open pavilion.

236. This early-nineteenth-century engraving shows a fragile Polynesian dwelling, wide open to sea and land breezes. Nukushiva Island in the Marquesas Archipelago. From Langsdorff's Voyages and Travels.

237. *The old-fashioned Japanese house comes equipped with a repertory of bamboo screens that drain the blazing sun of its heat and soften its glare.*

the tropical huts of Oceania, if considerably daintier. Indeed, in its summery deshabillé the Japanese house betrays its Pacific islands ancestry.

The architectural immodesty of these see-through habitations charmed explorers and shocked missionaries. James Cook, the discoverer of new worlds, was among the first Western men to set foot in the villages of the South Pacific. "The houses and dwellings of these People," he wrote apropos the Tahitians, "are admirably calculated for the continual warmth of the Climate; they do not build them in Towns and Villages, but separate each from the others, and always in the Woods, and are without walls, so that the air, cooled by the Shade of the Trees, has free access in whatever direction it happens to blow."[1]

Vernacular architecture of torrid zones is rich in ingenious cooling devices that have not been noticed, still less investigated. Until a few years ago the roofs of Hyderabad Sind, Pakistan, bristled with *bad-gir*, spooky-looking contraptions that conferred to the town its characteristic skyline. Their purpose was to scoop up the cool afternoon breeze and to channel it into every recess of the multistoried houses. Today, bad-gir are on the wane, giving way to electric fans. Admittedly, the wind sometimes dies down too early in the day, but so does with regularity the electric current.

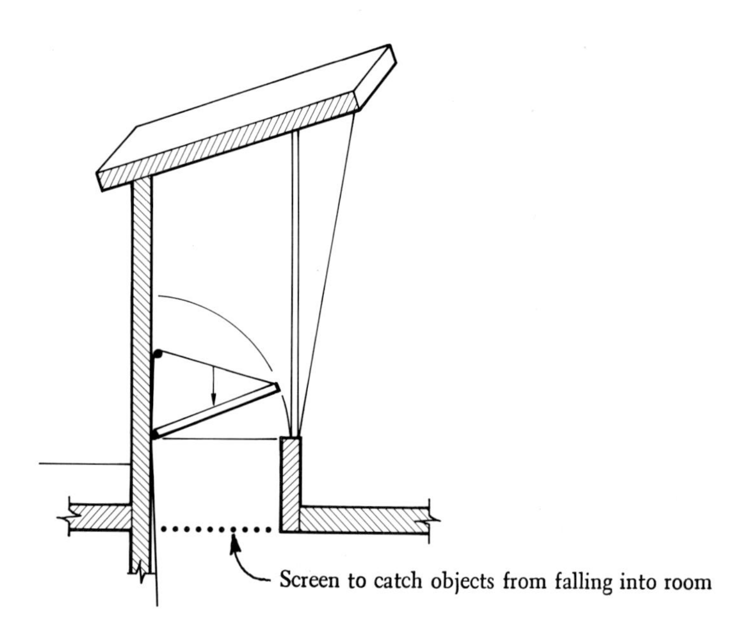

238. *Above: At Hyderabad Sind, in Pakistan, windscoops have been in use for at least five hundred years. Since the breezes that cool the town always blow from the same direction, the windscoops are fixed. Each room is ventilated individually. (Courtesy Professor Harold R. Benson)*

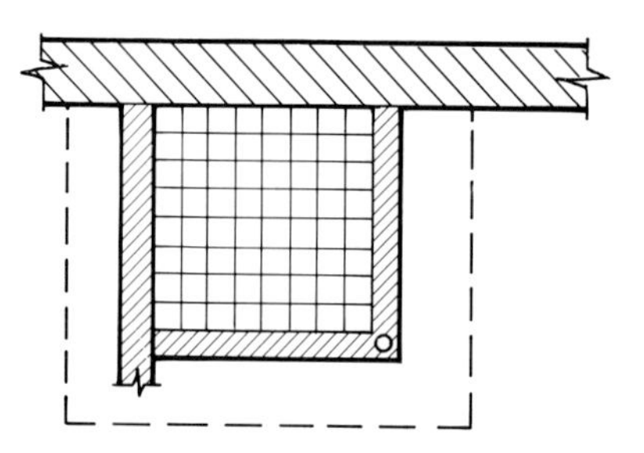

239. *Right: Although the Sindi now use concrete and metal in the construction of windscoops, their shape has not changed. (Courtesy Zaheer Alam and Mian Abdul Majid)*

240. *Opposite page: Windcatchers at Hyderabad Sind, Pakistan. From Martin Hürlimann's* India. *(Courtesy Atlantis Verlag)*

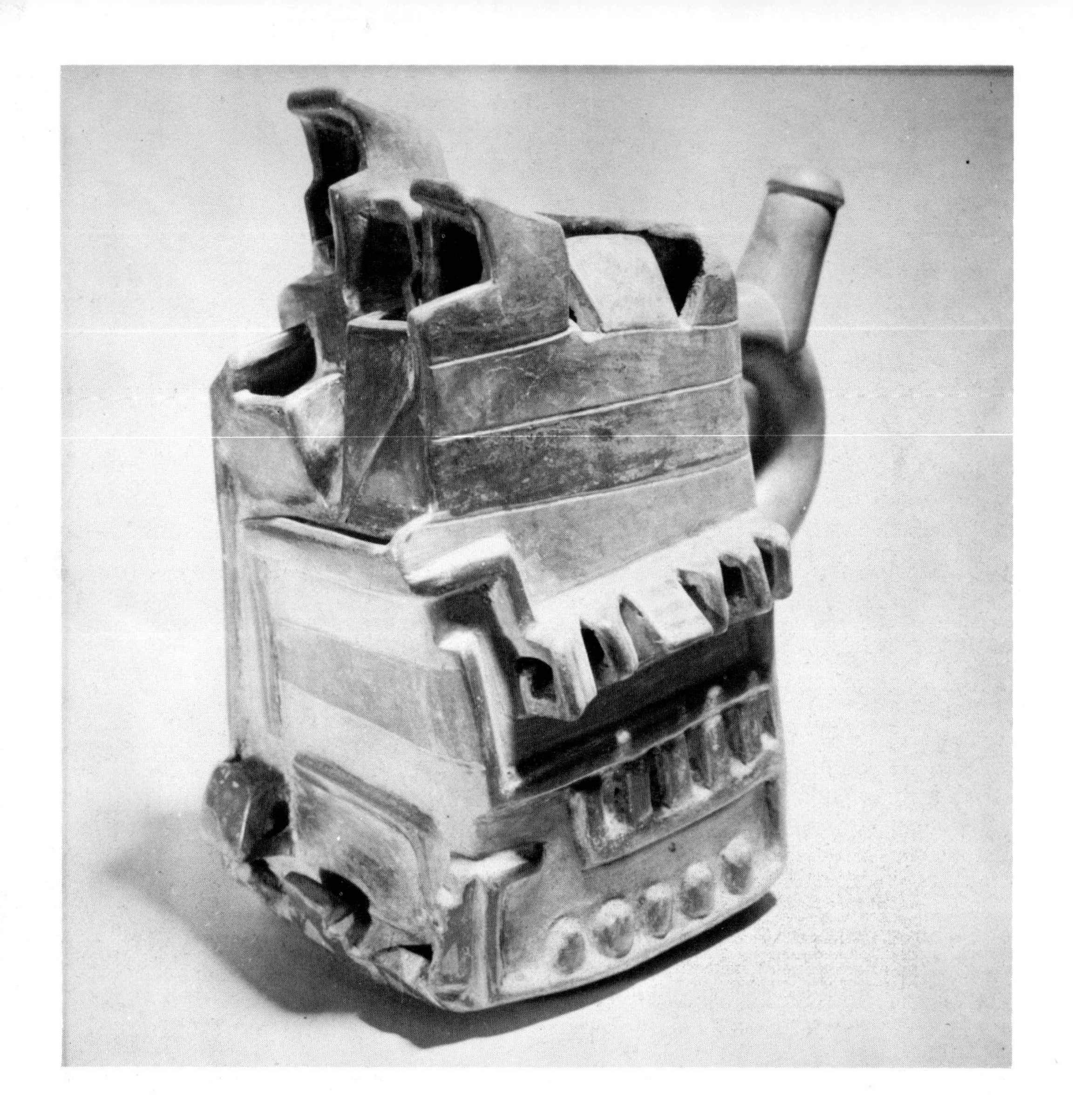

241. *The pot representing a three-story house is shaped in the naturalistic tradition of Peru's Mochica ceramics* (200 B.C.–A.D. 700). *While it may not be an accurate scale model, the several windscoops on the roof are most likely true to nature.* (*Courtesy* V. *von Hagen; Weidenfeld and Nicolson*)

Wind catchers are not uniquely East Indian; ages ago Peruvian Indians made use of them, as a model of a Mochica house shows. "Ventilation was usually acquired through vents on either side of the house," explains Victor von Hagen; "this invention of the window towards the sea, as with portholes on ships to re-direct air-wind [*sic*] into the room, is one of the interesting features of their domestic architecture."[2] It worked well enough to be retained by the Peruvians in colonial and republican times. Demonstrably, the oldest of these and similar contrivances is the already-mentioned *mulgaf*, or *malkaf*. An angular excrescence, resembling nothing recognizably Egyptian, mulgafs add acute accents and inverted circumflexes to the cuboid houses (figs. 242 and 243).

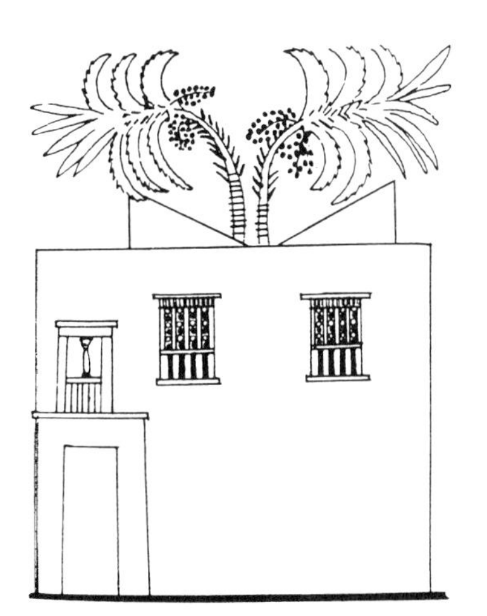

242. *Above: This papyrus depicts the so-called House of Nakht, an Egyptian country seat, implied by the date palm and sycamore. The superstructures on the roof are windscoops, similar to the old-fashioned cowls on a ship's deck. (Courtesy British Museum)*

243. *Left: Windscoops set back-to-back on the two-storied house of Nebamun. Middle Kingdom. (Courtesy Hassan Fathy)*

Climate does not necessarily dictate a specific roof type, and neither does topography. The general assumption that roofs mimic the contours of the landscape—more especially, that their pitch increases in mountainous regions—is disproved by the prevalence of flat-roofed villages in the Atlas and the Himalayas. Another popular misbelief attributes moral superiority to people who live under a steep roof. In the Third Reich the pitch of a roof carried political and racial connotations. The flat roof was tabooed in domestic architecture, its sympathizers branded as degenerates.

In warm climates the flat roof plays a vital role. It collects the precious rainwater which often is the only water there is. A full cistern makes a lifeline. On the roof tomatoes are spread in shallow wooden trays to soak up extra sunshine before being thickened into

244. Houses with flat, earth-covered roofs are characteristic of villages in the North African mountains. After the conquest of Spain, the Berbers introduced this dwelling type in Andalusia. See figure 245.

paste. In the fall the harvested grapes, prior to being pressed, are sun-sweetened for a few days. But mainly, the roof is turned to good account as an outdoor family parlor. On summer evenings and far into the fall people congregate there to stay until all hours. From the roof news is broadcast viva voce. Shouting from rooftops is standard practice; "what you talk about in your chambers, will be preached from the roofs" (Luke 12:3). Much to the urbanite's surprise none of the roofs have railings. Children run about or perch precariously at the edge, and by no means under their mother's vigilant eye. Anyone familiar with building codes meant to safeguard against accidents watches the scene with disbelief. His apprehension is unfounded. A child with his wits about him, he is told, is no more in danger of falling off the roof than a cat or a dog.

245. Pampaneira is one of a dozen near-identical villages in the Sierra Nevada with an architecture that goes back to the time of the Moors' occupation of Spain. The roofs of stone and beaten earth provide good insulation against cold and heat and serve as a place for work.

246. Lack of flat roofs induced the Venetians to erect wooden platforms on top of their houses where women once spent entire days sun-bleaching their hair to the proper shade of Titian-blond. Detail of Bellini's Procession in Piazza San Marco *(1496).*

247. Wooden decks on Venice's sloping roofs are still conspicuous in our day.

The pretty picture has its dark side. Thc village Montagues and Capulets are every bit as rancorous as their urban counterparts. Enmities smolder for years, indeed, for generations, and roofs are apt to become the theater of a phantasmal war of attrition. Walls are the duelists' chosen weapons.

To avenge an insult from his next-door neighbor, the wronged man will put up a wall to deprive the offender's roof of sunshine. The loser, determined to regain his place under the sun, dips deeply into his purse to add another floor to his house. No sooner done, the evil wall grows higher to restore the imbalance. *Architettura di dispetto*, spite architecture, bore its finest fruits in the Middle Ages when nobility vied with each other in erecting spiteful walls and towers.

The few that are still standing have long been promoted to landmarks, dead branches on the tree of architecture but far from being ruins. The rural kind, on the other hand, has lost none of its vitality. It thrives, unchecked by building ordinances. Yet only rarely does one of the rivals attain undisputed victory. I recall such a case from past days: One summer, when all but one of Anacapri's cisterns had run dry, the enemy of the fortunate owner saw his chance to get even with him. It was the choice of his stratagem that put him high on the list of imaginative *felones-de-se*: He drowned himself in the well-filled cistern where, pollutingly, he lay in state, fulfilled and triumphant in death.

248, 249. *As can be seen in both photographs, the sole purpose of the wall at left is to deprive the neighbor of his view and, more importantly, his quota of sunlight, indispensable for drying fruit, vegetables, and laundry.*

Next to the practicable roof, many a so-called primitive house's most valuable asset is the patio. Much misunderstood in non-Islamic and non-Latin countries, it is, as a rule, of modest size, more like a shaft than a yard, to keep out the midday sun. Nevertheless, it is anything but gloomy; scrupulously whitewashed walls transform it into a luminous well while a marble or tile pavement, generously sprinkled with water, exudes coolness. So does a small jungle of potted plants. Except for being open to the sky, the patio has little in common with its Californian namesake.

Patios are ubiquitous in North African towns where the space taken up by them by far exceeds that of streets and alleys. Bypassed by progress, these inner courts have preserved the purity of a cloister, minus its monastic severity. Uncompromisingly intimate, jealously guarded against intruders, the patio represents, so to speak, domesticity in the nude. It is a place for working and, occasionally, for sleeping outdoors.

250. In the palm-studded square of an Andalusian town the domestic awning finds application on a grand scale.

To keep a *large* patio from heating up requires more than a couple of beach umbrellas. What it needs is a sun sail. A first cousin of the

251. *At the height of summer a shopping street in Seville is transformed into a cool, dark lane by unfurling sun sails from the top of the houses.*

252. *A vertical sail closes the end of the street.*

253. *Here sacking is used as sun sail above the courtyard of a Spanish farmhouse.*

tent, it survives, precariously, in southern latitudes. In classical antiquity sun sails formed the roofs of circuses and stadia. They stretched over the Colosseum's bleachers, they provided shady promenades along the Via Sacra. During the following fifteen centuries, they served to cover a town's streets and squares, sheltering processions, audiences of mystery plays, auto-da-fés, and executions staged on dog days. It is not quite clear why they fell into disfavor. Hot countries may have cooled off, clothes certainly lost weight, and, on the whole, people nowadays prefer to make merry indoors.

In Europe, sun sails are making a last stand in southern Spain where they form extensive canopies between houses and across inner courts. In Seville, where "the climate is one of the most delightful on the continent" and the summer temperature rises to 115 degrees Fahrenheit, *toldos*—the Spanish word for awnings—cover the entire length of shopping streets. They run on ropes or wires, which makes furling and unfurling them as easy as pulling a curtain.

What toldos are to the Spaniard, the *pergola* is to the Italian. The corresponding English terms—arbor, bower, covered way—are helpful but neither exhaust its meaning nor give an adequate idea of the pergola's many applications. Unlike an awning, a pergola is a three-dimensional affair, if ever so incorporeal—a pavilion without walls and ceiling. It stands at the threshold of architecture as Nature's lobby, or, to turn the simile around, as the farthest penetration of architecture into the realm of Nature. Strictly speaking, it is an agricultural feature, a crutch for creepers.

254. *Wine pergola from the 1476 edition of Aesop's* Fables.

255. The leafy ceiling which spans a vast patio dispenses shade and coolness in summer and obligingly drops off in time, allowing the winter sun to enter. Granada, Spain.

256. Multistoried, 40-foot-high lemon pergole climb the shores of Lake Garda in northern Italy. During the cold season they are converted into greenhouses.

Raising a pergola is a joint undertaking of man and nature; it is planted rather than planned, with roots for foundations. The vines and calabashes, wisterias and ramblers are as much constituent parts of the structure as pillars and poles. Birds and butterflies supply the ever so volatile ornaments. No marble halls bear comparison to a fully orchestrated pergola with its vegetal arabesques, its sounds and scents. (The pergole here referred to are not to be confused with those pathetic contraptions by the same name—wooden trestles balancing on concrete beams, a standard fixture of hotel terraces and public gardens.)

Sun shades and pergole do not exhaust the repertory of non-machined refrigerants. There is the case of the Cypriote king who kept cool by surrounding himself with flocks of doves to fan him with their flutter. Where water is available, evaporation can be counted on as another means for lowering the temperature. Just as a wet towel thrown over one's shoulders brings relief from heat, so does wet matting or a wet rug hung over an open door or window. That eyesore, the earliest of automatic airconditioners—the box projecting over the gap left by a retired window pane—has been anticipated by a simpler and less ugly contraption: In India matting is mounted on a wooden framework that fits over a wall opening. By dowsing it with water, a tolerable indoor temperature is maintained on the hottest days.

So much for cooling. In temperate zones heating is of more immediate concern; in cold ones it is sometimes a matter of survival, and man does not hesitate to avail himself of that Stone Age amenity, an open fire. Yet long before he got around to building the first fireplace, he chanced upon a more dependable, more economic, not to say, more organic heating system—he tapped the radiant heat of animals. Indeed, the domestication of animals may very well have originated for reasons other than their food value and use as beasts of burden. A healthy animal emits considerable warmth while a herd amounts to a veritable heating plant. Hence in countries where fuel is scarce, family quarters are placed close to the stables. As a last resort, the stable itself becomes the living room.

Nothing could be more mistaken than to see an evidence of sloth in the presence of animals in human habitations. Only city people

257. In Armenia, where winters are severe and firewood is a luxury, living quarters are sometimes placed without intervening partitions between stables, whence radiates the aromatic warmth of beasts and accumulated dung. Although the stratagem may repel people addicted to air-conditioning, it suited Christ in the Manger. From Dubois de Montpéreux's Voyage autour du Caucase. (*Courtesy British Museum*)

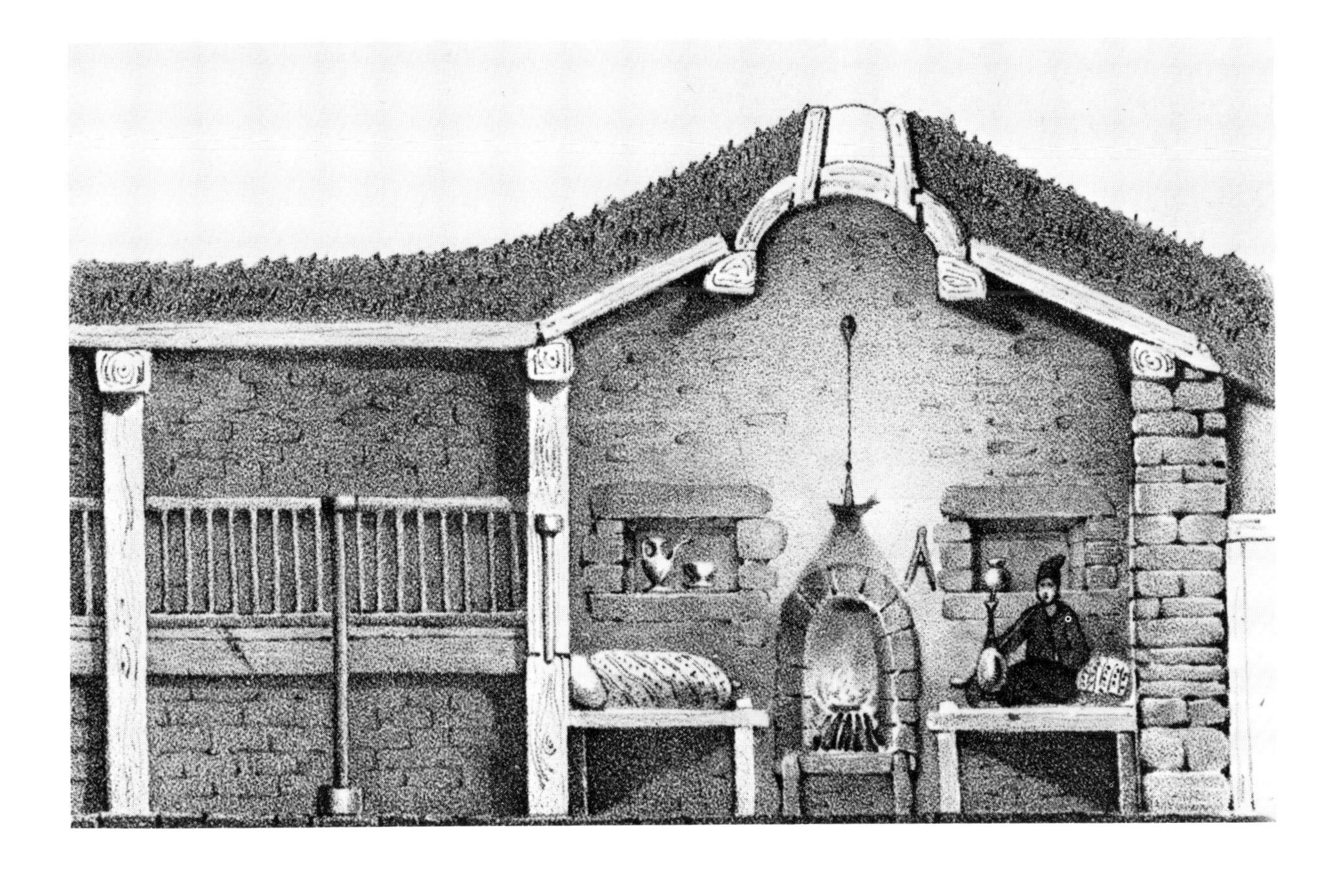

would regard the togetherness of man and domestic animals as sub-human living conditions. They forget that the beast as a fragrant radiator has divine sanction. According to Roman Catholic dogma, a stable was the Chirst child's first lodgings. Whereas we thaw our chilled bones by plugging an electric heater into a socket, Joseph simply pulled the Manger nearer to cow and donkey.

In its way, the fireplace—roasting as it does rather than warming us—is a far more primitive source of warmth than a drove of cattle. Our face may be feverishly hot while shivers run down the spine. Moreover, an open fire needs constant care, a state of affairs which

258. A monument to the culinary art of bygone days, this cubist stove, "a summer kitchen" from Curteny, Vaslui County, Rumania, has long since become a museum piece.

259. *A battery of rice cookers in the kitchen of a Japanese farmhouse, Yamato Prefecture.*

undermines the very fundamentals of an automated household. On the other hand, the fireplace addict will maintain quite rightly that the heat that radiates from the red-hot logs warms the cockles of his heart as much as his toes. At any rate, in the past the hearth stood for something more precious than a source of warmth.

That former role sometimes is still preserved in rural architecture. The kitchen stove, writes A. L. Sadler in *A Short History of Japanese Architecture*, is treated with deep respect as a source of sustenance for the house. It has a tutelary deity, like the well, the privy, and the garden. In the traditional house each hearth has several cooking stoves built into it. Some old country houses contain "one large stove that is unused and regarded as the sacred fireplace, while the work is done on a separate range."[3] Multiple stoves are nothing new; in the ruins of a Russian prehistoric village ten stoves in a row were found in one building alone, an arrangement one associates with restaurants only.

Much as we think of the oven as a late invention, it has a long history and an even longer prehistory. The distinction between heating oven, baking oven, and firing oven existed already in the neolithic age. Most, or all of them, were adjuncts to a well-appointed house. Their descendants are still around—the masonry-built tile stove, practically unknown in America but much in demand in European countries, and the immense Russian stove that lends the common room its peculiar character. The latter serves not merely for cooking and baking bread but as a social catalytic agent. Its top level forms a sort of upper berth for the nimble, while small children and the infirm occupy the surrounding wood or stone couches. To complete togetherness, small cattle and fowl are invited to snuggle up below. Surely, this monumental dispenser of warmth and companionableness merits a more fitting name than "stove" or "oven."

Another hybrid, the oven-bed, or bed-oven, is standard equipment in the old-fashioned houses and inns of Korea, Manchuria, and northern China. It consists of a platform of beaten earth, or paved with bricks, built along one side of a room and warmed from underneath. An institution of long standing, it is known as *k'ang*. A k'ang is illustrated in a tomb model of the Han period (206 B.C.–A.D. 220), but it probably came into use much earlier. The heat generates in a furnace placed in the room itself, or in an adjacent one, or, more commonly, out of doors. From the furnace it is piped into a space under the floor. Any fuel will do. Moreover, only a small fire is required to keep a house warm. The poor, reported a missionary, one Father Gramont, make the most of the firing that warms the k'ang. They sit on it by day—in the depth of winter this is the only way to cope with the cold indoors—and make it their bed during the night. The fire of the furnace also serves them to dress victuals, to heat their wine, and to supply them with hot water for making tea. "We should rejoice," wrote Gramont (in 1770) "if this hint could prove useful to the British nation."[4] His optimism was unfounded.

Scholars have debated the question whether the k'ang engendered the Roman radiant heating (the hypocaust) or vice versa. Floor heating probably originated independently in several countries. The system, best known to us from Islamic and Roman bathhouses, finds application also in Asian and African houses. Egyptians, forever at the mercy of extreme temperatures, do, or did, as the Chinese do.

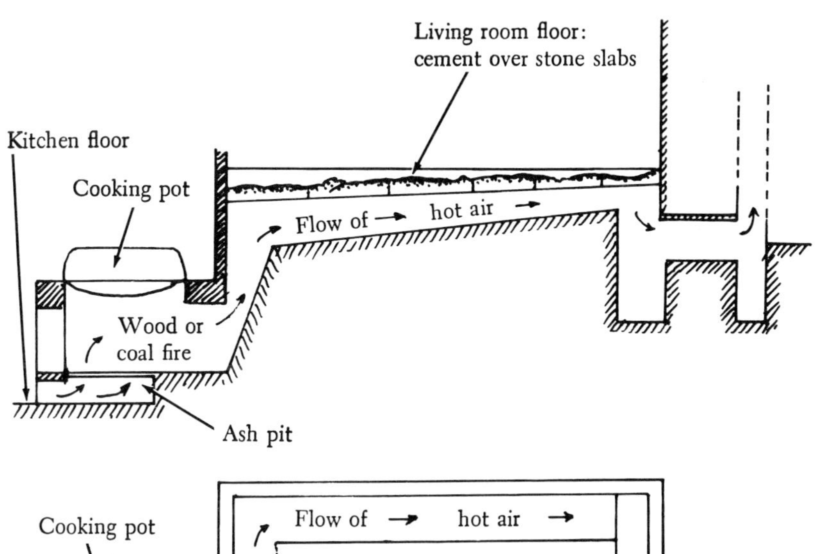

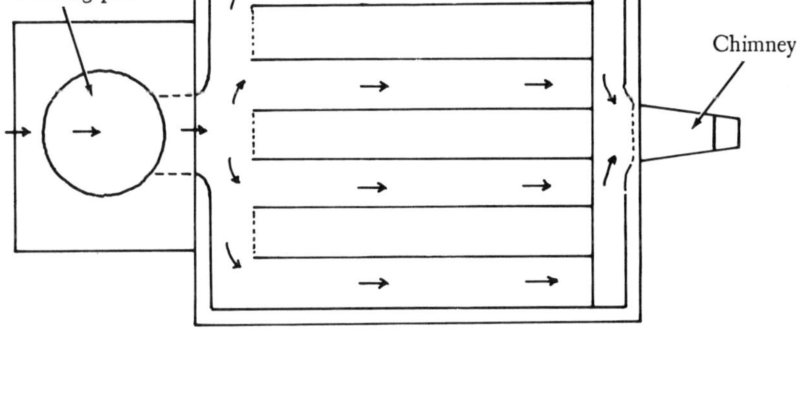

260. Floor heating, considered a luxury in our part of the world, has been a long-standing commodity in the peasant houses of Korea and northern China, where it is called a k'ang. *The plan and sectional drawing show the simplest setup for heating the living quarters, with the hot air generated in the kitchen stove.*

A source of warmth of a very different nature is the Japanese *kotatsu,* unknown and unsuspected in our parts of the globe. It consists of a charcoal brazier set into a recess in the floor. A table placed above it is covered with a kind of steamer rug, big enough to be pulled over the lower half of those who sit around it. Abiding by the adage Cool Head, Warm Feet, the Japanese use it as a sort of hot-air hip bath. By accommodating several pairs of legs, it promotes togetherness like no other piece of furniture excepting the bed. However, as the *Britannica* correctly points out, "the use of the kotatsu has disadvantages, often causing skin diseases" (fig. 261).

261. Kotatsu. From Japans arkitektur *by Svend Hvass. (Courtesy Royal Danish Academy of Fine Arts)*

262, 263. These clay models of sitting furniture were excavated at the prehistoric settlement of Yassa Tepe in Bulgaria. The picture of the chair (fig. 263, right) from Prehistoric Societies *by Clark and Piggot. Figure 262 represents the model of a sofa. (Courtesy Archaeological Museum, Plovdiv, Bulgaria)*

Did the early houses have seating comfort, so-called? Did prehistoric society loll on padded couches and upholstered easy chairs? Alas, some did. One would think that they knew better than let form-fitting furniture ruin their back. Instead, we have proof that our furniture-conditioned languor has early precedents. On the site of a prehistoric settlement at present-day Karanova in Bulgaria, models of furniture were found—benches, chairs, and couches—which even in their clumsy translation into clay strike one as products from the workshop of an accomplished cabinetmaker. Not a few of the ancient designs could pass for current ones (figs. 262 and 263). The low seats and low backs, combined with plenty of room for exaggeratedly long thighs, conjure up a Stone Age occupant with the sovereign sprawl of the tired businessman.

No such coddling was tolerated in northern Europe. Pre-Scotsmen, no doubt anticipating the mortification of the flesh perpetrated in latter-day public schools, stuck to hard, backless seats. At Skara Brae, the well-preserved Stone Age village in the Orkneys, a complete line of stone furniture was discovered in one of the houses. Apart from stone seats, there are stone slabs on either side of a hearth, "exactly similar to the fixed beds of planks built in the nineteenth century in Norwegian peasant houses, that on the right side being always larger. Tall uprights of stone at the corners like bed posts served to support some sort of canopy."[5] Equally remarkable is the presence of a two-tiered piece of furniture, "precisely like a modern dresser."[6] It is not quite clear whether the two passages, excerpted from an encyclopedia, stress the up-to-dateness of Stone Age carpenters or the backwardness of today's furniture makers.

264. This masonry bench built onto a Spanish peasant's house has the hallmark of antiquity.

The most calamitous event in the history of the house was its invasion and subsequent occupation by the various breeds of sitting furniture. For a hundred thousand years humanity had been getting along perfectly well without them. Not only Orientals and the rugged inhabitants of untamed continents sat or crouched on the floor, but Europeans, too, and by no means poor people, were able to survive, their dignity intact, without the unwieldy crutches. Before the opening of Japan to the West and the subsequent discovery of its "paper house" domesticity, it was the serenity of Turkish interiors that intrigued open-minded Westerners. "There was," reported in 1741 the botanist Pitton de Tournefort from a trip to the Levant, "neither bed, couch, bench, nor chair to be seen (for the Turks, of all people in the world, encumber a room the least with moveables) when at once a slave drew out of a cupboard in the wall all the materials for making our beds. When we rose, the same slave folded up the baggage in a moment, and put it back into the cupboard; and all this was done as swiftly as one can shift the decoration of an opera."[7] It is still done the same way in the traditional Japanese house.

The concept of the living room as a pure vessel of domesticity—airy, empty, and much on the manly side—in various Oriental countries lasted almost into our time, only to be discredited by the inexorable advance of industrialization. The house ended up as a storeroom for household goods, and never recovered from that lowly function. With conspicuous display increasingly assuming a social obligation, the inmate retreated to the position of a barely tolerated accessory, ending up as just another item in the house's inventory. The luxury of untrammeled space was for princes and paupers only, while the middling classes got mired in the swamp of so-called home furnishings. During the early phases of that movement that ushered in "modern" architecture, its protagonists attempted, if not altogether to liberate people from the tyranny of the house's ballast, at least to instill in their minds something like a yearning for space with a higher aesthetic oxygen content. Their efforts were not rewarded with success. Although they did weed out some of the Victorian underbrush, they fell short of exterminating the thicket of chairs.

Sitting, reclining, and lying are by no means clearly defined by our language. "Sitting" implies that our legs are hanging down, although there is no reason why one should not sit cross-legged on a chair. We sit "down" on a chair but sit "up" in bed. Posture is less a matter of muscles and joints than of etiquette. National custom alone is tolerated. Reclining at a meal, for instance, to Puritans smacks of debauch, despite the example of Jesus Christ and his contemporaries.

In the Balkans and points east, in old-fashioned households the dining table appears only at meal time. Portable and demountable, it consists of a tabletop and a folding rack no more than a span high. When not in use, the top leans against, or hangs from, a wall. Clearly, such a low table obviates chairs. (I treasure half-a-century-old culinary memories of chairless sumptuous repasts in the Turkish countryside when food was served not on individual plates but on gigantic platters right on the floor or on carpets spread in the shade of trees. It was eaten with one's fingers and a piece of bread for a rake, a method that calls for a good deal more dexterity than eating with Chinese or Japanese eating sticks. We are not so lucky; to earn a meal we have to work for it. Part of our food, instead of being brought to the table ready to eat as in Eastern countries, has to be attacked with metal tools, cut up into pieces and, if need be, dissected, extracted, carved, boned, cracked, skinned, or peeled, all of which ought to be the cook's business. The clatter of eating utensils and the clutter of objects on the table are audible and visible evidence of our ill-advised ways of doing things.)

A feature of many Oriental houses that lifts their inhabitants' life style literally above ours is the so-called living platform. It once was common enough throughout Eurasia, from China to the Near East, which reached not very long ago all the way to the gates of Vienna. The dimensions of this platform varied in time and place but the idea behind it was the same. It created an island of perfect cleanliness such as the West had never known or cared to know. Figure 265 shows a European example, the dais of an early nineteenth-century Greek house. The part of the living room where one sat is raised one step above the house's general level. Yet the few inches make all the difference since only the lower part was walked upon—in slippers,

265. Opposite page: Before their liberation (in the 1820s) and subsequent westernization, the Greeks were devoted to an Oriental life style of scrupulous cleanliness in uncluttered houses. The elevated living room with a continuous dais that served as divan was out-of-bounds for boots and shoes. The view of this early-nineteenth-century interior is from Trachten und Gebräuche der Neugriechen *by C. M. Stackelberg (1831).*

that is. As in Japan, street shoes were not tolerated indoors. Although the neo-Greeks in the picture were diehard Europeans, they had sense enough to see merit in the Turkish oppressors' custom. Yet as soon as the country was liberated and made safe for European civilization, they forgot all about uncalled-for footwear and furniture, gave up their regal bearing, and took to chairs.

This domestic dais has long been supposed to have originated in the extreme Orient. At the time of the Han dynasty the Chinese used to sit or recline on it, either unsupported or propped up by portable armrests. Only in recent years has evidence been unearthed which proves that the indoor platform is much older than had been assumed; the excavations of prehistoric Çatal Hüyük brought to light rooms containing two to five platforms. "These platforms, as carefully plastered as the rest of the house and frequently provided with round kerbs," writes James Mellaart, the excavator of the Anatolian town, "are the prototypes of the Turkish sofa (and divan) and served for sitting, working and sleeping. They are often covered with reed or rush matting as a base for cushions, textiles and bedding. Below these platforms the dead lay buried. . . ."[8]

Çatal Hüyük is noteworthy for other peculiarities. The houses had no outside doors, which was just as well since the town had no streets; the way to enter a house was through a hatch in the roof. (Pueblo Indians arrived independently at a similar expedient.) Moreover, doors between rooms were no taller than 30 inches, or about the

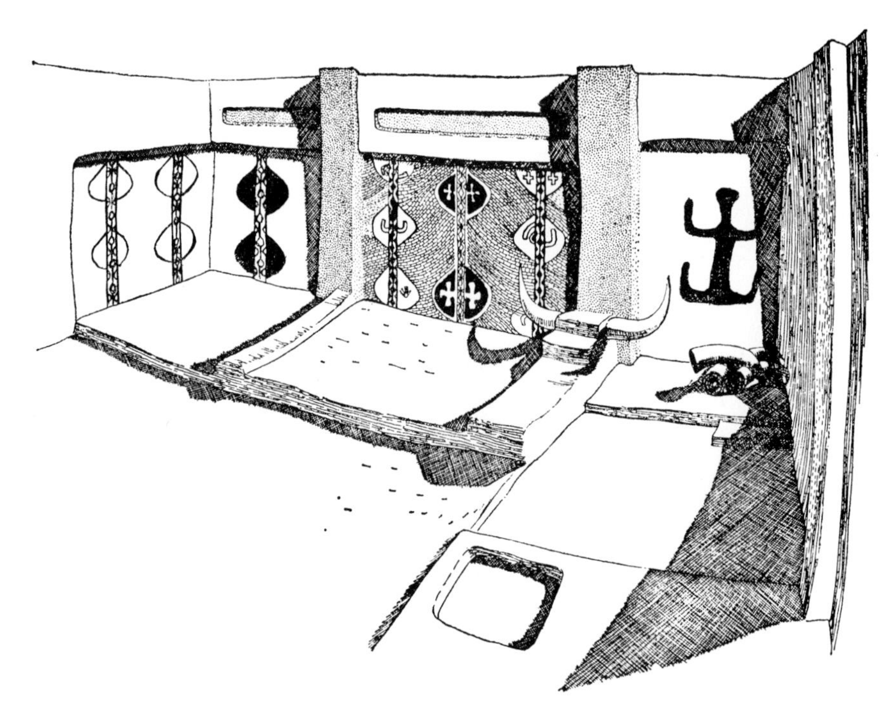

266. Built-in furniture preceded movable furniture. In Anatolian Çatal Hüyük, beds, couches, benches, and shelves were integral parts of architecture as early as eight thousand years ago. Excavations of the town's shrines and houses disclosed wall paintings and religious images and symbols of protection such as bucrania, the horns of the wild bull. The plastered sleeping platforms were covered with cushions and bedding. From Çatal Hüyük *by James Mellaart.*

height of the opening through which one crawls into a classical Japanese tearoom. It probably never occurred to the inhabitants that one might want to walk through a door erect. For all we know, a 30-inch-high door was considered preferable to a 6-foot one.

Predilection for minimal doors was widespread in prehistoric times. The porthole in the side of a ship had its forerunner in the hole, round or oval, that Stone Age man cut into a stone slab. The round type was small, just big enough for a person to wriggle through, but since it usually served as an entrance to a burial chamber, there was not much traffic and little inconvenience caused. Oval openings were considerably larger and made fewer demands on acrobatics. They occur mostly in sanctuaries and may have had a ritual significance which escapes us. Their exaggeratedly high sill is no great drawback since lifting one's feet high is less bothersome than jackknifing oneself into a low-slung car.

Before doors were deemed indispensable for a well-built house there was just the doorway, and before that, the cave's mouth. Uninvited beasts and irksome neighbors probably made a caveman feel the need for some kind of lid. To roll a boulder into place every time he left home was not the answer. A hide slung across the opening, though it did not prevent passage, at least stood proxy for a door. The arrangement is less wretched than one might think. It is still used in southern European countries where in summer the entrance to a church is often to be found behind a heavy leather curtain (gratuity to the sexton for lifting it) that takes up the entire width and height of the monumental door frame.

But Stone Age man not only felt a strong affinity for stone; it was sometimes the only available raw material from which to make his tools, weapons, furniture, and idols. It also served as canvas for his paintings. Stone doors, then, were the logical choice, and they must have proved their worth since they fell out of use only centuries after Christ. (Those impregnable fortresses, Scottish brochs, presumably were equipped with stone doors revolving on stone pivots.)

Like the orifices of the human body, doors and windows maintain the lifeline between indoors and outdoors. Windows, particularly, are indispensable for admitting light and air. Or so one might be inclined to think. But what seems natural to the layman runs counter to one of the professional's most cherished articles of faith. "Often I used to gloat over the beautiful buildings I could build," confessed Frank Lloyd Wright, "if only it were unnecessary to cut windows into them."[9] Although few architects are as candid as Wright, they share his view; wall openings to them are a calamity. Often they are less interested to provide their fellowmen with suitable shelter than to pander to their aesthetic prejudices. They pine for the ancient for-

267. *Left: Although the primeval doorway—two upright megaliths carrying a third one as lintel—were something of a hallmark of Malta's prehistoric architecture, the ancient builders did not flinch from hollowing out a man-size opening in a slab of stone.*

268. *Right: The humble garden entrance to a Morrocan house invites comparison with a moon gate, the circular wall opening in Chinese architecture.*

269. Door and window of this house in the oasis of Walata in Mauretania are typical of the traditional way for improving the looks of the austere domestic architecture.

tresses with those eyeless walls that form such immaculate cubes and cylinders.

Windows have been a headache to architects since remotest times. In Canaanite mythology, Baal, the god of fertility and life, commissions the divine craftsman Kothar-and-Hasis to build a palace to cover 24,000 acres. Its size poses no problem, but patron and builder disagree on whether it ought to have a window. It just happened that at the time a new type of building, the "window house" came into vogue. Against Baal's objections, Kothar-and-Hasis introduces the novelty, with the dire result that Mot, the god of sterility and death, enters the window. The event is echoed by the Prophet who specialized in bad news: "For death is come into our window, and is entered into our palaces, to cut off the children from without, and the young men from the street" (Jeremiah 9:21).

In the East—Near, Middle, and Far—the window conflict, or what to architects seems to be a conflict, spawned a variety of solutions that lift wall openings out of the commonplace assigned to them by us. Instead of leaving a hole in the wall, however cleverly styled and adorned, masons from Byzantium to Delhi resorted to a kind of light filter. A screen made from stone or wood not only is perfectly adequate for keeping the inhabitant in touch with the outer world, it also does away with shutters, blinds, and curtains.

270. *Fireproof shutters of a storehouse in Kawagoe, Japan.*

271. Sutomi, *an old Japanese device for closing a wide opening with a horizontally hinged panel, the prototype of our garage door.*

272. *Above: Shutters of horizontally hinged windows (and doors) also serve as awnings in Oriental domestic architecture. The typical Turkish house in Kavalla is a relic from the time when Greece was part of the Ottoman Empire.*

273. *Archaic Spanish demidoor, an ancestor of the Dutch door.*

The window's use as observation post found universal acceptance, but nowhere has it led to more ingenious solutions than in vernacular architecture. Spying on one's fellow creatures seems to fill a deep-felt need. To see and not be seen while taking in the street scene is a prime requisite for cultivating neighborliness. The *musharrabieh* of old Cairo, the wooden gazebo that to Westerners conjures up the sultriness of a Thousand and One Nights, has its counterpart in other countries where woman is jealously hidden from the evil eye of man. The precautionary window has many variations, but rarely has watchfulness been carried to greater lengths than in Spain. Although it is a long time since Muslims ruled Andalusia, windows of Arabic derivation are still much in evidence.

274. Most Orientals differ from Westerners in their conception of a house's windows; theirs is the complete antithesis of the blatant picture window. A deep-felt need for privacy and a partiality for filtered light spawned a variety of window grilles. India's aristocratic edition is a lace curtain translated into marble. Red Fort, Delhi.

275. *The* musharrabieh, *or* mashrabiyya, *is a screen made of turned wood that sometimes takes up an entire wall. It diffuses the harsh light and at the same time provides a sort of soft-focus view of the outdoors. (Courtesy Hassan Fathy)*

A special claim to architectural distinction can be made for the window that steps out of the wall, so to speak, into the third dimension. It ranges in extent from the size of a soap box to an alcove and is a boon to people with a penchant for neatly circumscribed cubical spaces, besides availing them an ample field of vision. (Its genteel version is our bay window, whose aesthetic shortcomings are implied by the colloquial use of the word for paunch.) In the Balkans, where Asian residues abound, the gazebo can be seen on peasant and burgher houses, but nowhere is it as conspicuous as on Malta. Unlike the rustic Andalusian species, on Malta it is an integral part of the high vernacular. Projecting windows that run continuously for a hundred feet or more along a building's façade are nothing out of the ordinary.

276. To the Spaniards, a window, whether of a farmhouse, townhouse, or palace, constitutes not just a wall opening but outdoor space, usually enclosed by an iron grill, which serves as a combination of balcony, alcove, and greenhouse. The sill supports flowerpots and idle elbows.

277. Enclosed balconies are a distinctive feature of Malta's High Vernacular. At Valetta, the island's capital, they nearly eclipse the houses' façades.

278. *Old-fashioned Spanish windows are equipped with lateral panes for casting looks out of the corner of one's eye.*

279. *Overbred window grilles of the High Vernacular at San Fernando, Spain.*

Spain's vernacular developed wall openings ranging from the trifling to the grandiloquent. The tiniest of them are vertical slits of a palm's width or less, set into the window embrasure—like the firing slits in archaic military architecture—which expose the outside to optical enfilade. To make these supplementary eye holes doubly effective, a kind of auricles are scooped out of the masonry (fig. 280). Parodoxically, the sensation of being watched increases in inverse proportion to a window's size. Whereas the reflecting panes of a porte-fenêtre turn sightless eyes to the passer-by, the lateral peephole follows him with a fixed stare.

280. In small towns where vehicular traffic has not yet driven out the milling crowd, people still prefer people-watching from a window to the entertainment proffered by the quivering screen. The scallops in the wall permit a wide-angle view. Arcos de la Frontera, Spain.

281. Ribbon windows, the hallmark of architecture in the 1930s, are native to northern Spain.

Finally, there is the case of the window triumphant that merges with the neighboring windows right and left, above and below, to cover acres of wall expanse. A spectacular feature, characteristic for northern Spain, it is endemic in Vitoria and La Coruña. Particularly in the latter town, glass-enclosed balconies, called *miradores*—literally, spectators—have long set a style. The waterfront with its glazed houses aglitter, mirroring sky and sea, have earned La Coruña the epithet "crystal city." Yet any similarity to our curtain walls is accidental; the Avenida de la Marina is no Park Avenue. The Spanish houses are eminently inhabitable; unlike our sealed glass walls, the miradores' windows can be opened.

282. Galicians built curtain walls long before they were introduced into modern architecture. La Coruña, Spain.

The call of the Minotaur

One need not resort to mathematical theorems or paradoxical conclusions of philosophy on architectural space to arrive at a sympathetic understanding of its nature. Conspicuous space configurations abound in their raw state as mountain gorges, mountain circuses, volcanic and meteoric craters. On a lesser scale, exquisitely molded enclosures are sometimes formed by chasms, coves, and dune hollows, above all, by those amphitheaters that result from cultivating the earth. The terraced fields of the Far East are good examples of what lately has been dubbed earth architecture.

But one does not have to beat a path to the Orient to form an opinion of this gratuitous quasi-architecture. Europe, too, has its share in the vineyards that climb like Jacob's ladders the steep shores of Rhone, Rhine, and Douro; the hanging gardens of Lake Garda that rival those of Semiramis; not to forget all those diminutive plots that have been wrested from mountain slopes in centuries-long travail. However, it is in stone quarries that man-made outdoor space reaches its apogee.

Call them earthworks or landscape architecture, quarries do not follow a builder's blueprint; they are the waste product of an industry, the separation of stone from its bed. Although some of them attain heroic scale, they have not been noticed, still less appreciated, if only because they do not carry the burden of history. Besides, the most famous ones, the marble quarries of antiquity, are the least attractive, gravitating as they do toward chaos. The ancient quarries of Pentelic

283. *Opposite page: Horticulturists occasionally join the ranks of inspired builders. These lemon gardens, now partly abandoned, rise perpendicularly from the shores of Lake Garda.*

marble, for instance, that supplied the raw material for the works of Praxiteles and Phidias, are without volumetric content, so to speak. They look like open wounds in the sides of Mount Pentelikon. It is the lowly quarry, the source of ordinary building stone, which fires the imagination.

Unlike marble quarries, which are mostly found high up in the mountains, the commonplace quarry is at home in flat country. Where people like their houses to be everlasting, towns usually have their local supply. On Malta, for instance, an island famous for its building stone, active and retired quarries stretch for miles—gigantic pits, framed by perpendicular walls of creamy color. Under the intoxicating sunlight, architects get lyrical about the soft limestone and compare it to honey and halvah. But then, a good deal of architecture, and not the meanest, has been likened to victuals. (By contrast, the sludge of frosty hues and sour smells that tumbles from the concrete mixer more likely induces nausea.)

Not until one stands at the quarry's edge does it reveal itself. One would think that its vastness destined it for uncommon uses such as a rallying point for conspirators; the headquarters of a holy sect or

284. The residual rock formations of an Italian quarry mimic the shapes and texture of buildings.

285. Among some of the most attractive of man-made spaces are the fortuitous rock formations which result from quarrying. They form a connecting link between primeval landscape and purposeful construction.

286. A grotto in the Latomia del Paradiso near Syracuse, Sicily, the quarry which in antiquity served as camp for Athenian prisoners of war.

highwaymen; a wartime hideout for townspeople; a trap for invaders; or the scene for a skirmish fought at close quarters. Instead it has been turned into a vegetable garden. Some quarries with perfect acoustics have the makings of a concert hall, yet commercial enterprise failed to recognize them as potential auditoria. The so-called Ear of Dionysius, for example, a crevice in a retired quarry, the Latomia del Paradiso near Syracuse, acts as a formidable amplifier of sound, without an interfering echo. It is supposed to have been used as a rehearsal room for actors of the adjacent theater, the largest one in

Greece. Its name, however, derives from the belief that it enabled the local tyrant, Dionysius, to eavesdrop on the whispered conversations of his prisoners. The more-than-100-foot-deep Latomia (*latomia* is the Latin word for quarry) made history as the most notorious concentration camp of antiquity. In 413 B.C., seven thousand Athenian prisoners languished for eight months in this monumental jail before being sold into slavery. Today it is covered by luxuriant vegetation, its pretzel paths filled with camera-clicking tourists who are given the shudders less by its lurid past than by its delicious coolness. Yet all the quarries in the world would not amount to more than a footnote on architecture were it not that they represent a link with a mythical building type, forever undiscoverable, though spiritually everlasting: the labyrinth. (The subject in question has nothing to do with the topiary mazes of rococo gardens; it refers solely to maliciously twisted underground space.)

A *labyrinth* is defined as a structure consisting of a number of intercommunicating passages arranged in bewildering complexity through which it is difficult or impossible to find one's way without guidance. The single most terrifying ingredient of the mythical labyrinth is darkness. Although dark spaces carry us right down to the realm of the dead, darkness is far from being a destroyer of space. On the contrary, darkness expands it. Nothing could be more inaccurate

287. *These massive towers, 35 feet in circumference, stand in the Cave di Cusa, a quarry on Sicily's southern coast. They are nascent drums of Doric columns carved, though not separated, from the live rock. They were intended for the 8-miles'-distant Apollo Temple at Selinus.*

than Sigfried Giedion's contention that it is light alone which creates the sensation of space. "Space," he asserted, "is annihilated by darkness. Light and space are inseparable."[1] Which sounds like the credo of a photographer rather than that of a historian. Above all, it denies the blind his grasp of space when it is he who has the keenest perceptivity of it. To compensate for his missing sense of sight, he develops uncanny powers of orientation by acquiring antennae, so to speak, for receiving space sensations far more subtle than those of a seeing person. He is able to perceive space through interpreting sound reverberations and changes of temperature and air pressure. Spinning out, mentally, Ariadne's yard, he takes echo soundings, if mainly unconsciously. A batlike sensitivity permits him to negotiate obstacles in- and out-of-doors without mishap.

If labyrinths haven't received much credence it is because of their singular elusiveness; even those postmortem examiners of architecture, archaeologists, have only the foggiest notions about them. Ancient tales alone have preserved the memory of ingenious mazes—corridors, stairs, and murky chambers, moist with fear, that an overheated fantasy populated with dragons and minotaurs. To us this underworld sounds every bit as confusing as the subterranean passages and staircases that lead to the lavatories, hairdressers, and giftshops of our big hotels and railroad stations. Modern communication devices and easy-to-follow signposts save us from losing our way into and out of the most fearful of modern labyrinths, but they also are to blame for our disenchantment with architectural space.

LABYRINTHUS
HIC HABITAT
MINOTAURUS

288. Graffito on a house in Pompeii. The inscription "Here lives Minotaurus" probably is an unflattering allusion to the inhabitant.

A web of legends has been spread around the ancient nonplaces and the men and monsters that inhabited them. They were created to explain the unexplainable; to translate the unspeakable into stone. The notion that in the depth of labyrinths human beings were sacrificed to a taurine monster can probably be traced to garbled reports about Minoan entertainments where prisoners of war were thrown to the bulls. This may or may not have been the distant prototype of the present-day *torida*, the Spaniards' postprandial recreation. A less cruel variant, unambiguously depicted in Cretan murals, is that of female acrobats seizing the bull by the horns and turning a somersault, a feat that the most sanguine of bemedaled Olympic gymnasts would be loath to attempt. Still, it may have come natural to women brought up on a diet of honey and heroic tales. According to bullfighting traditions of bygone days the *salto del trascuerno*, the leap across the bull's head, was a quite common maneuver of Sevillan toreros. So was the *salto del testuz*, the leap along the bull's back. As for the *mujeres torero*, today's female bullfighters, they probably never heard of their Cretan sisters' accomplishments, and studiously stick to the rules of their calling.

289. The trajectory of a Minoan acrobat's somersault as conjectured by archaeologists.

The fable of the monster-slaying youth, who finds the palace's exit thanks to a length of thread obtained from the king's daughter, endured as the most popular of nurse's tales far into historic times. A faint remainder of *threading* one's way through a geometric maze can be made out with the scholar's help in the old English children's game of hopscotch. But true labyrinths were no playgrounds. The very real dangers of underground chambers are not to be underestimated; anybody stepping on a couple of scorpions in the darkness of an Etruscan tomb may get an idea of this architecture's potential as a death trap.

No trace has been found of the most famous of labyrinths, the one on Crete that brought together a star cast including King Minos, son of Zeus and Europa; his daughter Ariadne, later to become the heroine of many an opera; Theseus, slayer of the resident Minotaurus, half-man, half-bull, by now, thanks to Picasso's etchings, as familiar as Panda; and, last but not least, the spurious Daedalus, allegedly the labyrinth's architect. (Daedalus was commissioned to build the maze as a stable for hiding the royal family's shame, Minotaurus, the offspring of Minos' wife Pasiphaë and a white bull.)

Neither Homer, Hesiod, nor Herodotus mention the Cretan labyrinth. The experts' guesses concentrate on Knossos in whose fourteen-hundred-room palace flourished the cult of the double axe, or *labrys*, the Minoan religion's most sacred symbol. The un-Greek labrys of unknown origin is the very object into which Zeus' spiritual essence enters, much as the Host in the pyx represents the material shape of Christ. From *labrys* derives the term *labyrinth*—that is, the house of the double axe.

Dissenting scholars prefer to believe that the Cretan labyrinth was a cave rather than a building, thus leaving Daedalus high and dry.

290. Specimen of the labrys *or double axe, the Minoans' most sacred symbol. Archaeological Museum, Iraklion, Crete.*

291. *With the help of this 1821 map of the Labyrinth of Gortyn, a subterranean quarry in Crete, one can easily retrace Theseus' course from the Entrance Hall (at the bottom), past the Deceptive Alley, Ariadne's Chamber, and the Entrenchment Room, all the way to the scene of Minotaur's defeat.*

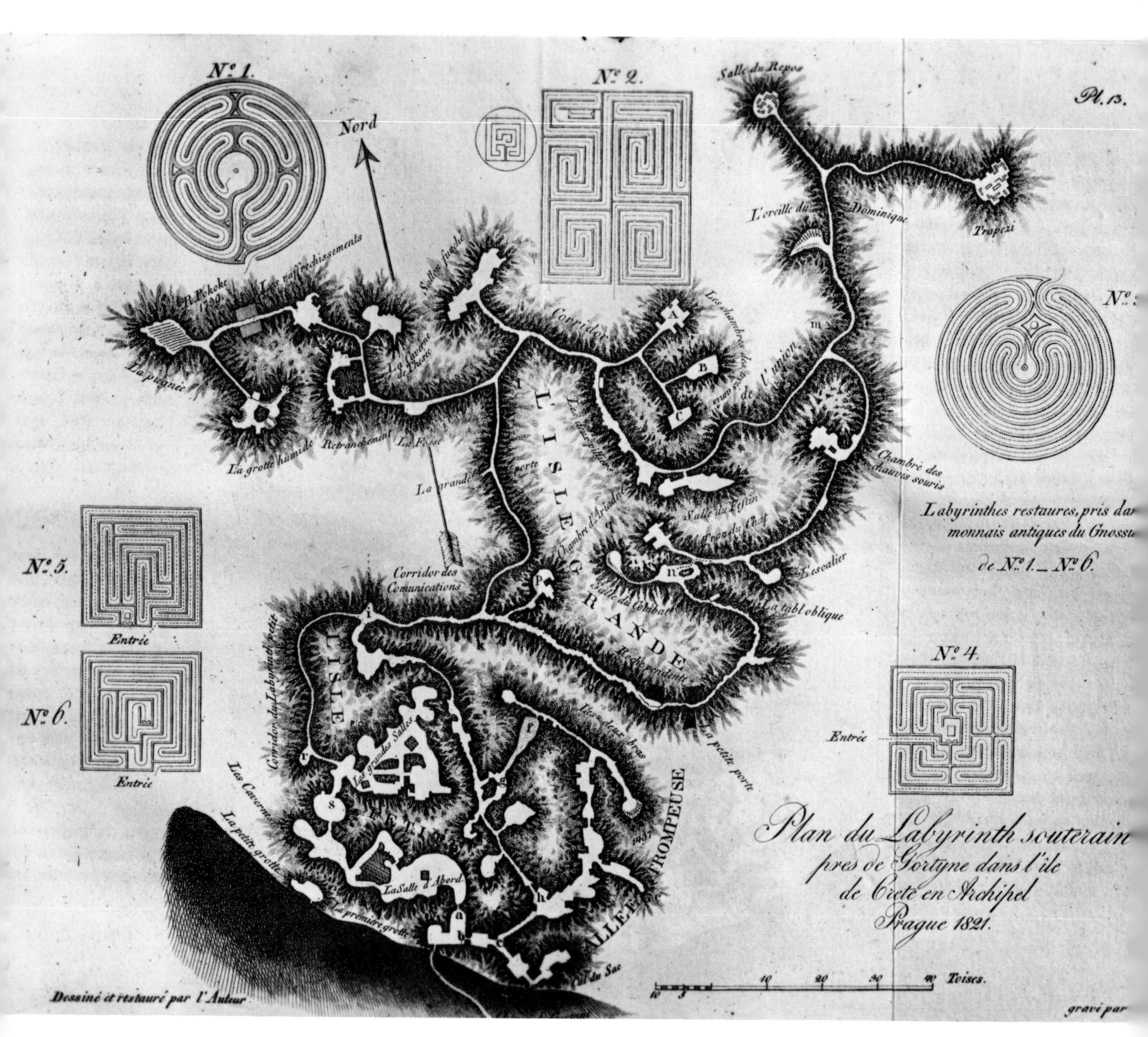

Hitching onto the opinions of the Roman poets Catullus and Claudianus, they assume that what gave rise to the romantic distortions of Minotaur and his spooky premises was the so-called Labyrinth of Gortyn. Be that as it may, Gortyn, according to Strabo, built by Minos, was next to Knossos the island's most powerful and prosperous city. As late as A.D. 1415 the Italian Buondelmonti, while visiting its ruins, counted no less than two thousand fallen statues. Obviously, the buildings of such an important city—it measured 6 miles in diameter—called for a lot of stone, and the quarry that furnished it must have been extensive. Its location has not been ascertained but there is, not far from Gortyn's ruins, a hill at the foot of Mount Ida that contains some remarkable cavities, traditionally associated with Gortyn's quarry as well as with the labyrinth.

The above-mentioned Pitton de Tournefort left a detailed description of the place. "As one advanced beyond the entrance," he wrote (in 1741), "all was full of surprise. A corridor of random course and irregular form led to two large and beautiful halls. However, one was constantly in danger of getting lost among the thousand alleys and narrow lanes. As a precaution each of us carried a big torch, and at the most tricky twistings we left numbered slips of paper and had our escorts put up markers of straw. We also posted several guards who would alert nearby villagers should we fail to return."[2] It is difficult to understand why the learned traveler did not avail himself of that most reliable expedient, Ariadne's thread, all the more as in those times thread was touched by something like divine ordinance, spinning being the occupation of ladies of noble birth.

At any rate, Tournefort thought it most unlikely that the Gortyn labyrinth ever had been a quarry, since the stone was neither attractive nor hard enough to serve as building material. "Why on earth," he asked, "would people want to look for stone at the end of a more than thousand foot long lane, crisscrossed by an infinity of others, in which one risked getting lost at every moment?"[3]

Other visitors came to different conclusions. Although they were to a man loath to put faith in subterranean chambers palmed off by the cicerone as "Ariadne's Room" or the "Hall of Combat"—that is, the scene of the fight between Theseus and Minotaurus—they did find indications of quarrying, such as square-hewn stone of considerable size lying around.[4]

Etymologically, the evidence does point to quarries as the prototype of the labyrinth. Not only was the labrys—the double axe—a divine symbol; in antiquity it served as the quarrier's tool. Quite probably, neither the Palace of the Labrys nor the quasi-quarry of Gortyn, least of all Herodotus' Egyptian labyrinth, ever existed in such shape as we like to think. For they share with Heaven and Hell

292. *By the time of the Renaissance the story of Theseus and Ariadne had been updated and enriched, the heroes outfitted to meet the taste of the day, and the labyrinth compressed into a cake mold. This depreciation of mysterious space into a dull architectural scheme further led to the creation of intricate pathways bordered by hedges, the gardener's version of the maze.*

the distinction of never having been visited by people who returned to tell about them. Indeed, all the creepy tales and specious maps that men piled up in the pursuit of labyrinths merely betray wishful thinking.

This elusiveness makes them all the more fascinating, more a-mazing. The concept of the artful maze transcends the mere physical; it goes back to a mythopoeic age. Hegel understood labyrinths as paths between incomprehensibly intertwined walls—not, as he said, with the silly (*läppisch*) proposition to find the exit but intended for "meaningful ambling among symbolic riddles."[5] Whatever its merits, the remembrance of mythical architecture ought to be treasured in our shabby present, for it may yet inspire future builders. For the moment it suffices to know that today's most grandiose would-be labyrinths are the end products of sober quarrying and mining rather than images evoked by if ever so attractive lunacy.

293. *Christianity never quite disowned the burden of paganism it inherited. Thus, in the Middle Ages the labyrinth turned up in places of worship, albeit relegated to the floor. To make it acceptable it was construed as an abstract Calvary; penitents seeking divine forgiveness negotiated its intricate course on their knees, an exercise that would have astonished Theseus. Octagonal labyrinth in the pavement of the cathedral of Amiens, France.*

One of these spectacular pseudoarchitectural spaces is the salt mine of Wieliczka in Poland, practically unknown in the Western world. A man-made maze of vertical and horizontal corridors running on seven different levels, it reaches down nearly 1,000 feet and has a combined total length of 65 miles. No ordinary mine, it once harbored a veritable town in its innards, hewn out of rock salt: churches, shrines, and many kinds of monuments; a central railroad station for a

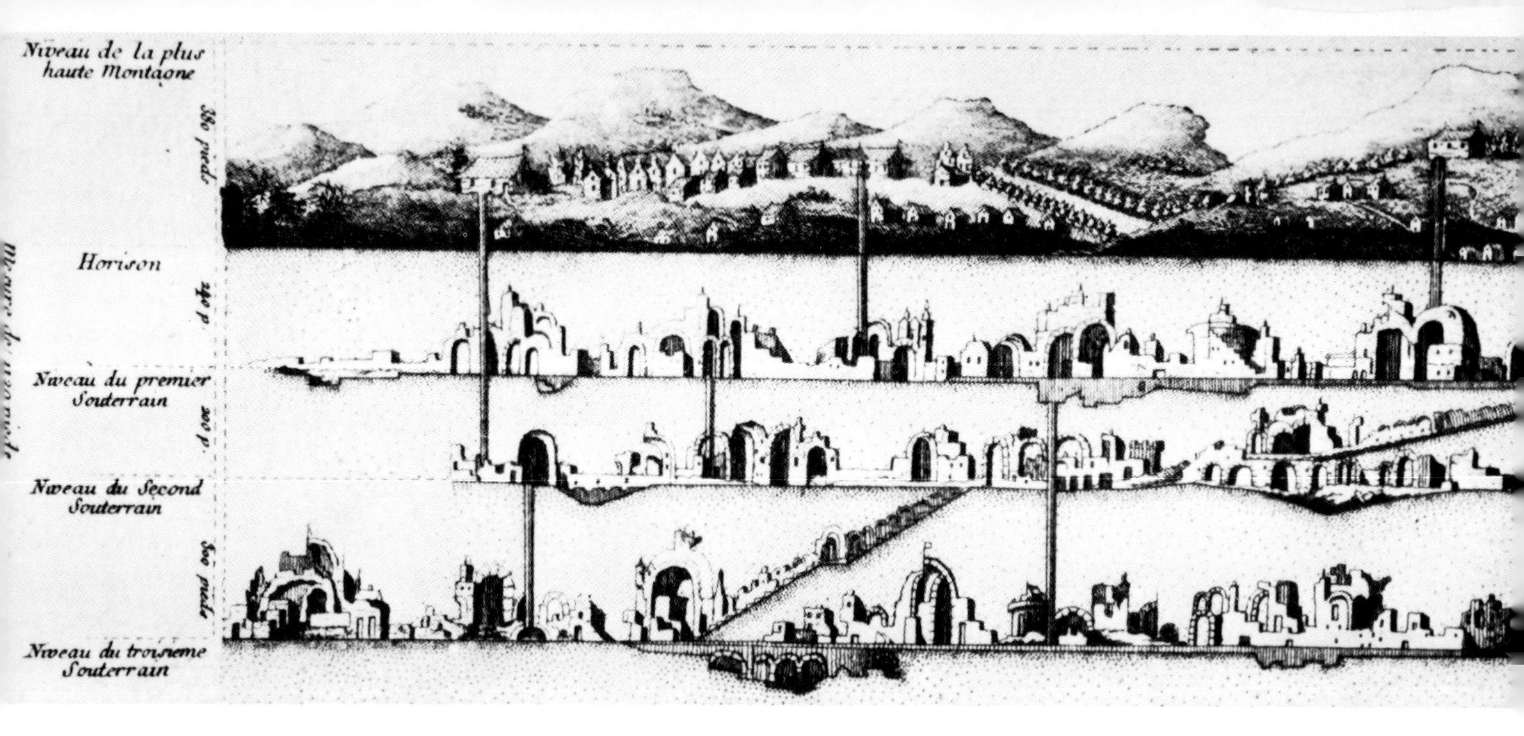

294. Among the most spectacular of man-made labyrinths is the salt mine of Wieliczka in Poland, which extends over 65 miles and has been worked since the eleventh century. Churches with elaborately carved altars and a ballroom 100 feet in height lend the mine some upper-world urbaneness. (Courtesy Biblioteka Jagiellonska, Cracow)

narrow-gauge railway; several restaurants and a dancing saloon, all accessible by monumental stairways. Sixteen ponds, the largest measuring 200 feet in length, and a clear rivulet of fresh water animated the salty Hades.

The mine's former vertical transportation system is worth squandering a paragraph. Where elevators are running today, in the past people were lowered on a rope into a narrow, pitch-dark shaft, hundreds of meters deep. Unlike on Athos, the rope was worked by a horse, and there was no basket to accommodate the rider. Instead, a man wanting to go down fastened a smaller rope to the big one and made a sort of loop into which he seated himself, with another man in his lap. He then signaled to be let down. "When several go down," related the historian Stephen Jones, who visited Wieliczka in the last years of the eighteenth century, "the custom is, that when the first is

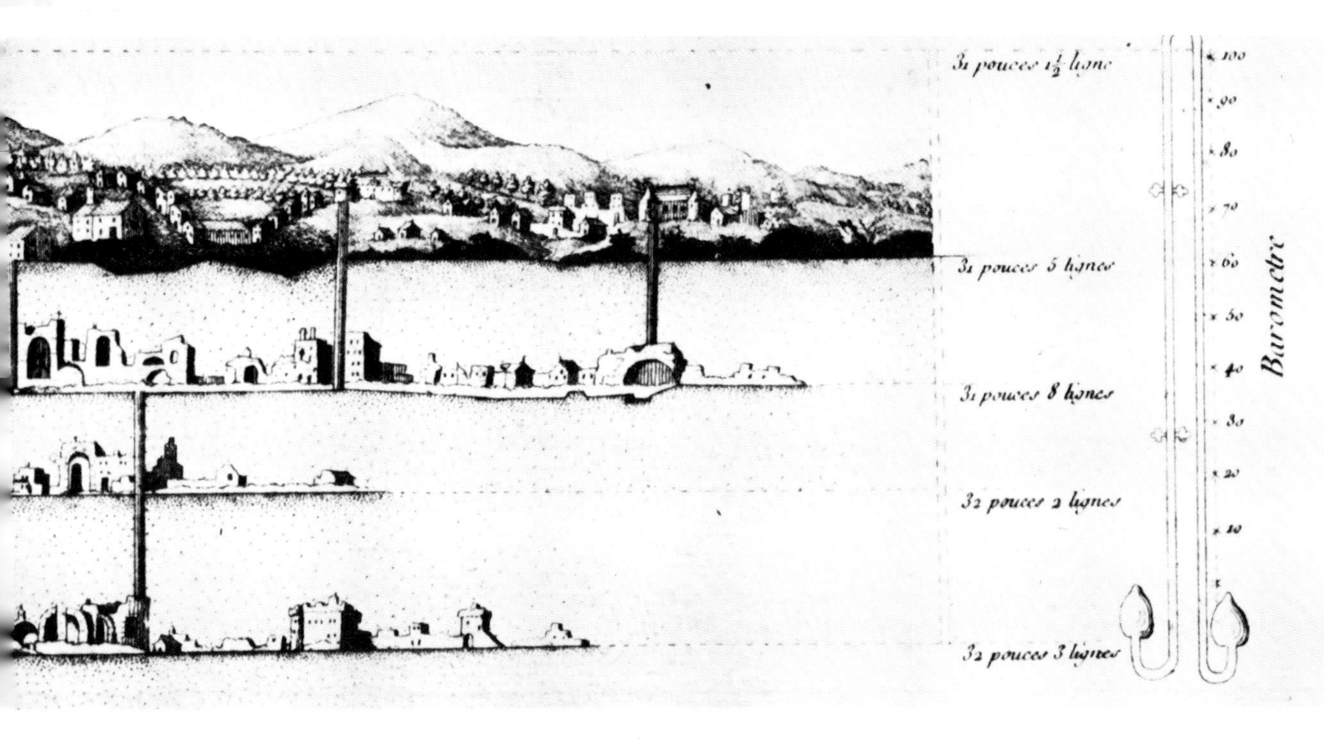

let down about three yards the wheel stops, and another miner takes another rope, ties himself, takes another in his lap, and descends about three yards further; the wheel then stops for another pair, and so on till the whole company is seated, when the wheel is again worked, and the whole string of adventurers are let down together. It is no uncommon thing for forty people to go down in that manner."[6] Speed was of no importance.

Now picture for yourself the pains that these people had to take to expedite horses in the very same way. To aid our imagination, an eighteenth-century engraver obliged with plates recording the transport of men and beasts in the shafts. Quaint and immoderately cumbersome though the old system may seem to the habitué of high-speed elevators, it is a pity that it has not been reintroduced for his delectation on holiday jaunts.

295. In the engraving can be made out, at left, an altar, its crucifix and candelabra carved from the sparkling mineral; at right, a stable and elevator for men and horses. (Courtesy Biblioteka Jagiellonska, Cracow)

296. Spotty lighting at Wieliczka's labyrinthine underworld discloses a swarming humanity at work. The solid wooden towers are for shoring up the gob. (*Courtesy Biblioteka Jagiellonska, Cracow*)

But back to Jones, who meanwhile has reached, if ever so slowly, the bottom of the mine where things are getting curiouser and curiouser. Terrorized by fear, he is being dragged by his guide through utter darkness when there bursts upon his view "a world, the lustre of which is scarcely to be imagined." The world that unfolds before his eyes is infinitely more wondrous than anything Alice beheld. "It is a spacious plain," Jones wrote, "containing a whole people, a kind of subterranean republic, with houses, carriages, roads, etc. This is wholly scooped out of the vast bed of salt, which is all a hard rock, as bright and glittering as crystal, and the whole space is formed of lofty arches, vaults, supported by columns of salt, and roofed and floored with the same, so that the columns, and indeed the whole fabric, seems composed of the purest crystal."[7]

The enchanted city in the bowels of the earth has all the ingredients of a fairy tale: the rabbit hole access; any amount of décor for the trial scene from the *Magic Flute,* here suitably transposed from fire and water to the salt of the earth; the cave from *One Thousand and One Nights* (to Jones the columns supporting the grandiose cavity looked like "masses of rubies, emeralds, and sapphires, darting a radiance which the eye can hardly bear"[8]) and to complete the improbable tableau, there were large numbers of people around, all as vocal as a well-rehearsed chorus of an Italian opera company. Instead of meeting human wrecks, Jones found the miners "all merry and singing." Not only were they at peace with their surroundings, they seemed to prefer their subterranean houses and villages to ordinary dwellings. What contributed to their happiness was the knowledge that they were less vulnerable to man-made disasters than the rest of the nation. During centuries, when Poland was ravaged by foreign

armies, life in the mines went on as usual, and the miners and their families survived unscathed. They have, wrote Jones, "very little communication with the world above grounds, and many hundreds of people are born and live all their lives here."[9]

The republic of merry trolls has long been dissolved and their territories turned into pasture grounds for tourist herds. In the 1950s, the mine was made into a museum. Its treasures are now spread over barely 2 miles of grottoes, including the chapel of the Polish princess and saint, Kinga, decorated with statues that may give the three hundred thousand annual visitors an idea of what Lot's saline daughters may have looked like.

Homage to the squatter

Travelers long on memory may remember the unredeemed Marcellus Theater in Rome at a time when its halls resounded to the din, not of first-night audiences promenading during the intermission, but of resident artisans, ambulant vendors and their customers, all of them congenitally noisy. Workshops then occupied the peripheral arches on the main floor that had partly sunk into the ground, or rather been buried by deposits of cultural debris, while the theater's central portion which once formed the orchestra was taken up (it still is) by the Palazzo Orsini, tenanted by a savings bank. Despite two thousand years of incursions into its expanses by plebeian and patrician squatters or perhaps because of them, the building was still in fairly good shape, except for the loss of the top floor, a minor casualty, everything considered. Erected by Caesar and Augustus, it had served one purpose or another down to our days. Pregnant with a great past, ostentatiously picturesque, looking like a Piranesi engraving blown up into a life-size *tableau vivant*, its sight sent aesthetic shivers down the spine of the pilgrim making the rounds of Rome's relics.

Today the halls echo emptiness. The tenants were turned out long ago and siphoned off to housing projects. The theater fell victim to a purge in the 1930s when Mussolini tore into the town's archaeological center and disemboweled it with the deliberateness of Jack the Ripper. Rome emerged from the enterprise enriched by several Pennsylvania Avenues but it lost out on other points. Hoary constructions that for centuries had swarmed with life and never stood vacant even in modern times either were razed or—as in the case of the Marcellus Theater—shorn of their accretions and downdated, so to speak. What resulted from the cleanup and architectural spare-parts surgery were dehumanized landmarks, strangled by iron fences and locked gates, with perhaps a ticket window and a board listing visiting hours. The operation was successful, the patient dead. What proved wrong was the diagnosis.

Two elements that are forever omitted from the charmed circle of rhetorical architecture require clarification: the illegitimate inhabitant, or squatter, and the parasitic ingrowth of an equally illegitimate subarchitecture. The two occur almost exclusively in old civilizations whose glut of architectural monuments is squatterdom's bonanza. The sturdier the building, the more attractive it is to people in search of a roof. With an ancient stone vault for a ceiling and walls several meters thick, the self-confidence of the person whom it shelters soars to heights unattainable to those who dwell between plaster partitions.

Far from being that lawless lazybones and penniless parasite who insinuates himself into the frayed fabric of society, the squatter who makes his home in the nooks of ruins can be compared to the honeybee. He may not connect ruins with glory, but his very choice attests to a certain fastidiousness. Whereas clergymen instinctively pounce on pagan temples with the avowed aim to convert them into churches, the populace displays a preference for the spectacular, if long silenced, centers of mundane entertainment, such as arenas and theaters of bygone days.

Unlike the inhabitants of shantytowns, the occupants of these antique homesteads rarely originate with people from the lowest walks of life. On the contrary, they constitute a steadfast class of craftsmen and tradesmen who can be depended upon to keep the premises in good repair for another couple of centuries. No hermits, they bring their families and their tools and, in some anticipatory mood of fatality, set up shop among the cool millenarian stones.

What starts out as a reconnoitering party in time solidifies into proprietorship. In some instances the squatter hasn't budged from his premises in two thousand years. By virtue of his fixed residence he becomes an honorary trustee and unofficial guardian of a landmark. What takes place is something akin to a chemical reaction; just as nature converts plants and trees into solid rock by a wearisome process of substituting their water content with mineral matter, so the squatter fossilizes, as it were, the usurped building to a point where it becomes immune, if not to foul weather, then, with luck to the inroads of commissions for the restoration of monuments.

Squatters proved a blessing to architecture; whenever they moved into an abandoned building, its life expectancy increased. The architectural host—say, an aqueduct, or public bath run dry—gains from their tenancy. In almost every case, the takeover entails salvation. Hence the relationship between rent-free tenant and impromptu shelter ought to be termed symbiotic rather than parasitic.

Squatterdom cuts across classes and castes. The nobility was never too noble to need much urging when it came to appropriating choice bits of classical architecture. By making them into strongholds, they often brought a feudal atmosphere to the stones. However, no other power pressed its claims on the grand monuments as persistently as did the Church. Few are the temples and thermae that did not receive baptism and therewith protection from the wreckers' fury. Twenty centuries are not enough time for the elements to bring to heel a well-constructed building; wherever destruction succeeds, it is by the hand of man.

Whereas the demolition of a modern structure produces nothing more substantial than dust and rubble, that of an ancient one often yielded enough material for building a small town. Ruins, therefore, have been traditionally used as self-service quarries. The best known of them and the most productive single fountainhead of reach-me-down travertine blocks, today reduced to barely a third of its original bulk, is Rome's Colosseum. Only its belated consecration to the memory of the men and women who died in the cause of entertaining a hard-to-please audience stopped the centuries-long plunder.

Squatting poses major problems when indulged in by crowds. A town's entire population in flight from the enemy is less concerned with scouting for a roof over their heads than with finding a hiding place. In former times this was provided by that near-indestructible shelter, the houses of the dead. This is a reversal of the expedient of making a dwelling also serve as a tomb. A case in point is the necropolis of Pantalica at the southern tip of Sicily in the canyonlike Anapo Valley, a landscape quite uncharacteristic of Italy. For miles not the ghost of a house or hut, no olive or fruit tree, no field or forest can be seen. Neither is there a road. At the bottom of the chasm runs a mere goat path along a rivulet hidden by oleander bushes. Devoid of human scale and human traces, the countryside exudes a distinctly mythological aroma. At its northern end, where the valley abruptly ends into a kind of Pyreneean circus, the sheer cliffs of the mountain are dotted with thousands of man-made cavities which are the entrances to tombs cut out of live rock in prehistoric times. The city that supplied the dead disappeared long ago; only the tombs are left. In the Middle Ages the tombs proved their habitableness when they absorbed the population of a nearby town. The fugitives probably took the transfer in their stride with equanimity. Keen on being on good terms with ancestors, if ever so remote, they thought nothing of

297. An example of the interchangeability of tomb and habitation is supplied by the necropolis of Pantalica near Syracuse in southern Sicily. The thousands of burial chambers, carved out of the cliffs of the Anapo Valley by the prehistoric Siculi, were converted into multistoried dwellings in the Middle Ages.

moving in with them in times of distress. There are worse things than setting up housekeeping in a tomb.

At Pantalica space was anything but tight, and the horizontal and vertical passages that connect the funeral chambers engendered the sort of neighborliness that goes with multistoried apartment houses. The voices of children and the lowing of cows extolled domesticity, while the odor of death yielded to the fragrant smells of frying onions and peppers. In front of the doors unfurled a panorama of such dreamlike peacefulness as one could ever hope to find on earth. The lack of windows was not felt as a calamity because in southern countries the light has a tendency to wander around corners.

A less cryptic squatter community, shown in the accompanying engraving, is the Roman amphitheater at Arles, built in the second century after Christ. It might have suffered the fate of similar piles of stones by being quarried into nothingness had not a security-minded populace decided to put it to more profitable use. Surely, one did not have to be a town planner to grasp its suitability for accommodating a small town. With its two stories of sixty bays solidly walled up—as in the Marcellus Theater, the third floor was missing—it would constitute a formidable civic fortress in the round. And so it did. Only in this instance the squatters did not stop at usurping the spacious old structures but, turning builders for a change, inserted among them a settlement all of their own. Three towers, added in the twelfth century, did not enhance the architectural pastiche but lent it a mildly martial note. To the occupants' chagrin, the idyl did not last. In the 1830s they were evicted, the medieval barnacles removed, and the remaining fabric reconstituted as a ruin. The result looks just like dozens of other Roman amphitheaters.

It is not quite clear what purpose was served by the squatters' expulsion and the subsequent exhumation of a badly mauled building that was not likely to excite one's curiosity, still less one's admiration. An amphitheater is not much of a building to begin with. It is petrified landscaping rather than a temple of the Muses—an artificial incline for spectators watching spectator sports. Moreover, even one-and-a-half centuries ago, arenas had had their day. Christians were not anymore for burning, and the Spaniards' use of the arena as a stylish abattoir never gained favor with the French.

Had the squatters been left alone, today Arles would boast a rare example of a compact town within a town. Too compact, the urbanite will say whose beau ideal is the empty lot. Admittedly, medieval streets *are* narrow, yet they seem ample compared to the crevices formed by our high-rises. Since the Arlésiens did not have to cope with double-parking and air pollution, they could afford to keep their streets to the width of corridors without endangering their own dig-

298. Arles's second-century amphitheater once was the site of a full-fledged town. In the Middle Ages the Arlésiens, prospecting for urban Lebensraum, *usurped the massive Roman structure and turned it into dwellings. The theater's third floor was taken down to furnish the building material for the houses, churches, and chapels that fill the arena. From* Histoire d'Arles *by J. F. Noble de la Lauzière. (Courtesy New York Public Library)*

nity and health. So do to this day the inhabitants of Islamic cities or, for that matter, European towns of medieval vintage, like Siena or Perugia. However, urban hygiene—an amenity we seem to think we can do without—here is not the issue. The question raised by the disappearance of Arles's amphitheatrical inner city is whether, and how well, mankind is served by the indiscriminate restoration of monuments.

While the amphitheaters of antiquity get a full treatment in textbooks, bullrings, their direct descendants are omitted. To be sure, bullfighting—bullbaiting would be a more fitting designation—today is pretty much confined to the Iberian Peninsula, yet to the Spaniard corridas are as essential as bread. Indeed, many a village that boasts its own bullring may not have an *horno*, a baker's oven. Whatever one's reservations about attending a production of death in the afternoon, the stage and its background deserve comment.

The favorite spectator sport of the Spanish nation has its roots in the arena of the ancients. The ruins of Roman amphitheaters at Cordoba, Toledo, Mérida, and others were the scene of the earliest performances, introduced to Spain by the Moors, in the twelfth century. Towns that lacked such architectural relics used their main squares instead, and some still do. In those times, bullfights, far from

299. The townfathers of Tarazona, the former residence of the Aragonese kings in the Spanish province of Zaragoza, probably anticipating the fate that befell amphitheaters like the one of Arles, in 1770 built a bullring, half-arena, half-apartment house. One hundred years later the arena was retired and the 192 boxes walled up to provide additional habitations.

being routine Sunday entertainments for the masses, were the climax of coronation festivities, royal weddings, and similar grand occasions. Moreover, during the Middle Ages, to challenge a bull was the prerogative of nobles; living up to their station, as many as ten knights would lose their lives in a single afternoon. Kings of the truly regal kind thought nothing of descending into the ring and giving battle to the pedigreed monsters. Thus, Charles V, Roman Emperor, King of Spain, endeared himself to his subjects by killing a bull with a lance, which was then the proper weapon in this dissymmetrical duel. In the eighteenth century, bullfighting received its present rules and ever since has been performed in specially constructed establishments modeled, roughly, on the Roman circus. Madrid's first great Plaza de Toros, built in 1743, signaled the conversion of an aristocratic sport into exhibitions of professionals.

Whereas big arenas are as banal as sports stadia, some small ones do not lack architectural distinction. A chance discovery at Tarazona, the Turiaso of the Romans, led me to a rare variant of this profane architecture. A cross between an arena and an apartment house, it is quite likely the only one of its kind in existence. Its hybrid character may have been dictated by civic prudence; the fate of so many vast buildings that fell prey to squatters probably induced the citizenry to legitimize a takeover from the outset. Be that as it may, the arrangement proved successful. It satisfied the aficionados of bullfighting and benefited the would-be inhabitants of the peripheral premises.

Tarazona, the former residence of Aragonese kings, is a small town in the province of Zaragoza. As so many Spanish towns with a glorious past and rich architectural heritage, it often figured as the setting of sumptuous fiestas which inevitably included bullfights. Yet while guidebooks are listing its cathedral's and churches' artistic treasures down to the last candlestick, they ignore the existence of the old bullring cum tenements.

This bullring—the country's oldest next to that of Ronda—was constructed in 1770 by the local guilds on property donated by the authorities. Its octagonal shape alone makes it the only one of its kind. Moreover, its seating radically departed from the typical plan. Instead of bleachers it provided genuine box seats. Arcaded loggias on each of the three upper floors offered a more or less advantageous view of the goings-on below. Hence, typologically, Tarazona's old bullring is less an arena than an open-air auditorium on the order of the *corral*, the theater at the time of Calderon and Shakespeare. Except that here, with neither an orchestra nor a proscenium, the boxes went around all the way.

In 1870, when the building had served for a full century in its dual capacity, the town's corridas were transferred to a new bullring, and

300. *Dwellings inserted into an old aqueduct's arches—honey in the comb—furnish a convincing example of the reuse of superannuated architecture at Evora, Portugal.*

the space that had been vacated by the boxes was made into additional habitations. This is the way it presents itself today, with the arcades walled up and a small window inserted in each bay. The arena itself remains deserted. Now, another hundred years later, the building is up for a new lease on its old life. The days of the tenants are numbered; in fact, they are going to meet the fate of common squatters. As soon as quarters are found for them, they will have to vacate the premises. A victim to the categorical imperative of change, the bullring is earmarked for restoration. Recharged as it eventually will become with the crowd's excitement over *corridas goyescas* and some such amusements, it may easily last another century.

Squatterdom at its most enduring is exemplified by the peaceful takeover of an imperial palace on the shores of the Adriatic Sea and its conversion to a regular town. The Dalmatian capital of Split, for thirteen centuries known as Spalato, arose from the residence that the aging Emperor Diocletian had built for himself as part of his retirement plan. When the hypochondriac ruler abdicated his sovereignty in A.D. 305—an unprecedented event in the annals of Roman history—a marble palace covering nine acres, as hermetic as that other regal retreat, the Escorial, was waiting for him. (Their dimensions are nearly identical.) Quadrangular in shape, enclosed by walls which on the seaside attain a height of 80 feet, it spelled security in a region that was right in the path of barbarian incursions. The odd compound —a small world of its own—had been laid out along the lines of a Roman military camp. Four gates led to two thoroughfares that

301. The uncontested occupation of a monumental building complex by squatters is exemplified by Split's Vieux Carré, formerly Diocletian's palace. Roman vestiges that can be made out in the aerial photograph are the town wall; the octagonal imperial mausoleum, now a cathedral (to the left of the campanile); and the peristyle (to its right).

302. *On outdoor café nestles in the roofless peristyle of Diocletian's palace.*

intersected each other at right angles, but there any similarity to a *castrum* ended. The imperial fortress was on the sybaritic side: a jumble of arcades, inner courts, gardens, and apartments glutted with architectural loot from Greece and Egypt. A mausoleum and a private temple dedicated to Jupiter provided spiritual succor. Anticipating Voltaire's dictum, the august tenant spent the last ten years of his life cultivating his garden which was fed by a 5-mile-long aqueduct that still supplies the town with drinking water. Upon his death, the palace was first turned into a textile factory and subsequently occupied by people seeking refuge from Avars, Tartars, Turks, and their likes.

In sixteen-and-a-half centuries the squatters' rights remained uncontested. The Church claimed the mausoleum and temple. The one became the town's cathedral, the other was transformed into a baptistery, St. John replacing Jupiter. A café now occupies the peristyle, and mercantile enterprise flourishes among the palatial vestiges without prejudice to either. The seawall, which in antiquity was washed by the waves, now borders a dusty quay and a jetty built on reclaimed land. Advocates for restoring hallowed sites could do no better than to give this unsightly accretion back to the sea.

303. *The palace's seawall, once bathed by the Adriatic Sea, today faces a traffic street.*

304. Virtuoso block play accomplished this breathtaking monument in which man's playfulness and imperviousness to good sense are fused into a full-length portrait of his building urge. Shorapoor Hills, Dekkan, India. From Cairns *by Meadows Taylor.*

Block lust

"In every condition of humanity, it is precisely play, and play alone that makes man complete," wrote Friedrich Schiller; "man plays only when he is in the full sense of the word a man, and *he is wholly Man when he is playing*"[1] (Schiller's italics). T. S. Eliot concurred with this assessment when he said that the separation of work and play destroys culture at the root.[2] Play is of course not a human prerogative; some animals indulge in it with equal gusto. Lion cubs wrestle in mock combat; kittens chase a ball, and chimpanzees do pole vaulting out of sheer exuberance. There is no lack of scholarly studies that have explored the meaning of their games. Such youthful play, it is thought, sharpens the animals' instincts. We are less informed about what play does to children or, for that matter, to adults. "The play of humans" said a candid *Britannica*, "is no more clearly understood than is that of lower forms."[3]

On the whole, modern man's pursuit of play, unlike that of animals, is rather perfunctory. We are incapable of doing things we were not taught, and playing is one of them. Hence our solicitude for instructing the child how to play. Whereas in so-called primitive societies a child's play is spontaneous, the urban kindergarten tot has to be coaxed into playing. Cooped up indoors most of the time, he depends largely on toys for his amusement. It is the kinder-gardener's job to fan his interest in toys, and the best way to do so is through organized play.

The child who does not play is a being to be distrusted, warned

Beaudelaire. Play supposedly prepares the child for a successful adulthood. Yet it is perfectly conceivable that in our ebbing civilization the reverse will happen; a good deal of our adult life may be spent in conditioning us for a successful juvenile senescence. With labor, both physical and mental, increasingly deferred to nonhuman agencies, we may very well end up with all play and no work. To assess this threat to our sanity, let us take a look at the insidious side of child's play.

Pedagogues tell us that play is anything but trivial; it has, they say, "great earnestness and deep significance."[4] To acquit themselves properly, they consult a vast body of literature that deals with the upbringing of their little charges. One way to make play attractive is to provide them with an arsenal of "educational toys." To keep up with the progress of education, toy manufacturers have long been sounding psychologists. According to the latter, there is nothing like that tried standby, blocks, to stimulate the senses. "We have been led to the conclusion," says the author of *The Art of Block Building*, an inspirational book of the 1930s, "that blocks are essentially the most admirable, plastic material for young children."[5]

(A-2) Tower Building

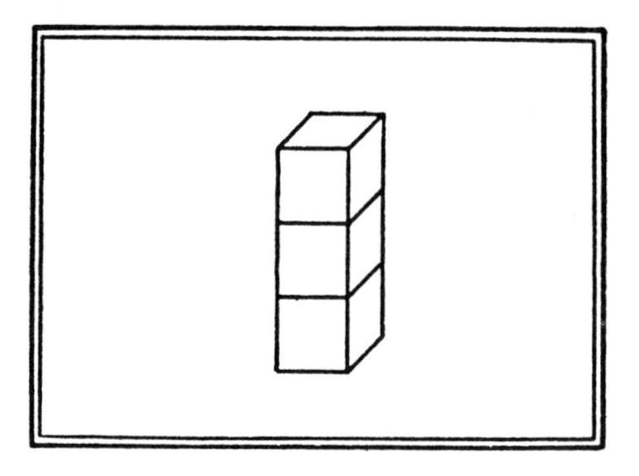

305. *Illustration from* The First Five Years of Life *by Arnold Gesell et al.*

This is a mouthful of a statement, yet psychologists are a singularly intrepid lot. "Blocks," noted one of them, "have the clear approval of the children themselves, since they use them more often than any other playthings."[6] Still, one cannot help suspecting that consumer preferences of the nursery child reflect the educators' tastes. Blocks, concurs a third, are "universal in their appeal to children's activity interest. It is the exceptional child who does not respond to them."[7] And what American parent likes the idea of nurturing an exceptional child under his roof! If one is to believe the experts, these blocks represent nothing less than chips from the philosopher's stone; "any school," one of them declares, "that pretends to an educational philosophy includes opportunity for block-play in some form."[8] Surely, there is more to playing with blocks than the uninitiated might think; the philosophical implications of the toy point straight to the realm of mental blocks.

(What about the *safety* factor in playing with blocks, asks the wary parent defensively, and forthwith falls into the salesman's patter: no prongs, no spikes, no catches or teeth. Obviously, he never met the self-reliant unblocked child, the proud custodian of his collection of pebbles, pods, seeds, gourds, and suchlike winsome trifles.)

Curiously, the bleak cube as plaything was not devised by some malevolent abstractionist of our time but by a mild-mannered, humorless, early-nineteenth-century German schoolteacher, Friedrich Fröbel, the originator of the kindergarten. If he had any tenderness for the young, he did not dissipate it on them. He arrived at his true vocation in a roundabout way by trying his hand at several professions. He served, successively, as forester, architectural apprentice, museum curator, and soldier in the Prussian army. Wartime duty taught him the merits of discipline, yet it was probably his bout with architecture that led to his infatuation with the cube. Convinced that toddlers' play is not just a way of passing time but the free expression of a child's soul, and desirous of engineering "spontaneous" play, he hit at the least likely object one could think of, the hexahedron.

His choice was never seriously contested, and henceforth geometric abstraction became an important ingredient in the nursery pabulum. "It is essential that the blocks be cut very accurately," expounds a latter-day block-playwright in her epic style, "so that all edges are even and that the multiples and divisions of the unit are exact, as they are tools for the children's use, and the most desirable building habits will be established only if the materials are stable and precise."[9] Such categorical statements probably make parents fear that a child playing with anything as coarse as pebbles or shells may turn into a dimwit. To be sure, the cube did not crystalize in the human mind; it occurs in nature as salt or fluorspat crystals, but one's chances of encountering them outside of mineralogical collections are negligible.

It comes as a relief to learn that although blocks have the approval of nonexceptional children, the other higher primates withhold their endorsement. One cannot say that they are unduly capricious or blinded by prejudice in the choice of toys, yet none of them share children's enthusiasm for playing with blocks. "Chimpanzees reared in isolation," writes one naturalist, "avoid even simple and entirely innocuous objects, such as a small block of wood or scrap of cloth, on first encounter. Indeed, an animal might require daily exposure extending over several weeks before it makes the first hesitant contact with such an object."[10]

Fröbel's choice of blocks was prophetic. At the time he received his call to mold the minds of the little ones, the better part of the world was still tidy and fragrant, abounding in forests and clear streams, with the sea's plangent waves unsullied by oily squish. However, in the course of time it was to undergo a fateful change that led to a new attitude toward life. It saw the destruction of the woods and the darkening of the waters, the growth of gigantic factories and block after block of equally gigantic housing projects. The infant had a preview of things to come right in the nursery, for his first encounter with the hard-bitten reality of life were hard-edge building blocks. And to this day they still are. At home and in schools (pretending to be an educational philosophy) on shelves within easy reach, the cubes are stacked in rows, allegedly for the child's amusement, actually for imprinting on his mind the principle icon of our architectural creed.

Let us see what the connections are between playing with blocks and the proliferation of the sort of architecture that for lack of a more precise epithet is called modern.

To judge from the textbooks, the child's love for the blocks is neither spontaneous nor instantaneous. At first he seems to share the chimpanzee's distrust, and not until the age of eight months does he begin to notice them. Twenty weeks later he places one block on top of another. If he does not conform to the educator's timetable, there are ways to speed him up. At the Yale Clinic of Child Development it has been found that a child can be induced to build a tower of blocks before he is ready to do so on his own. Or, if the child waits to be told what to do, say, "*Build something for me, anything you wish.*"[11] The advice and italics are Dr. Gesell's, the doyen of child psychologists.

If coercion is persistently applied, the child can be expected to have caught the builder's itch at the age of fifteen months. His style may be still diffuse, but at least he is able to ingratiate himself with his persuaders. "On his best performance"—at the age of three—"he can build a tower of ten cubes."[12] Beyond this, it seems, there is no further climb; the tower is the crowning achievement of nursery architecture. By then, we may presume, block lust has seized him; "*even the ten-year old does not scorn blocks*"[13] (italics added).

He is a dull parent who cannot see the rewarding side of block play. That side has been scrupulously studied and pondered upon by the doctors and recorded in the *Archives of Psychology*. "Chief use of blocks," writes F. Guanella ("Block Building Activities of Young Children"), "includes knocking down."[14] The parallel with some grown-ups' practices is evident: For the first time in architectural history men of business and architects plan buildings with their future demolition made easy. Built-in obsolescence can be relied upon to shorten the span between birth and death of a construction. Although a fallen tower of building blocks looks puny compared to the wreckage of urban renewal, the satisfaction derived from engineered collapse differs only in degree.

To give an example on the infant's level: "Carl," reports Ruth Hartley, the author of *Understanding Children's Play*, "is showing great delight in piling up the blocks and throwing them down. . . . He bangs a block with his fist very viciously, using a great amount of energy. Very softly while doing this, he says, 'f - - - you, f - - - you' "[15] (Miss Hartley's dashes). No doubt, we have here an instance of what art historians refer to as the education of the private ego. Creative originality, argues one of them, manifests itself only when it finds a problem that it can carry toward a solution. As if our educators had not known! The perspicacious Miss Hartley has grasped this point better than any city planner. "It may be," she conjectures, "that knocking over block structures has still other values. *It is a way of controlling the environment* and asserting possessiveness; What one creates, one has a right to destroy"[16] (italics added).

The child eagerly adopts this article of faith and tries to live up to it in later years. So much so that during the last generation the science of wrecking has far surpassed the science of building. Today, wrecking ranks as a special branch of technology. The objects of destruction are not just houses and towns but entire countries. Wrecking, civic and military, figures as the most expensive item in the national budget. Is it too farfetched to assume that the germ of blockbusting is planted in the kindergarten? Modern psychologists have found that a child's outlook does not change much in the process of growing up but persists unchanged, and eventually is absorbed into the mental background of the adult, "Le génie" said Baudelaire, "c'est l'enfance retrouvée." Here, childish destructiveness ripens into mature vandalism.

The problem child, deficient in *furor americanus* and disinclined to control the environment through creative tantrums, always gets a helping hand. To speed up destruction, his mentor-tormentor will egg him on to deliver the coup de grace to his work. "After having observed the limits of his spontaneity," proposed Gesell, "the examiner [*sic*] need only urge him to continue until the tower falls."[17] He hopes that, given enough time, our obsession with achievement will catch fire in the mind of the least aggressive child. The rise and fall of the pile does not exhaust the blocks's possibilities. According to the expert, they also include "noise-making manipulation: pounding and thumping, dropping, bouncing, tapping another child's head with a block, throwing, beating, dumping," not to mention "licking, chewing, sucking and biting" blocks.[18]

Yesterday's block-busting tots are today's overachievers—plotters and planners, chips off the Old Block—who are debating, if not solving, urbanistic questions of the many-layered malaise brought on by our architecture. Of course, today there is no unanimity in nursery schools on how to either agitate or pacify children. Some of them never get around to playing with blocks but are given instead a wide range of equipment to indulge in the more fashionable pastime of painting. The proponents of this kind of activity, however, received a nasty shock when news transpired in 1957 of a sensational one-ape show, so to speak, in London's stately Institute of Contemporary Arts, of paintings done by a chimpanzee. The work of the young artist which initiated "a new and biological approach to art,"[19] was in direct competition with the senior expressionists of the day. No nursery tot has been able to challenge the prodigious ape.

We have to cross the frontiers of our civilization into what until recently we used to call the world of savages to encounter children untouched by industrial toys and mechanized entertainment; whose curiosity has not been deadened by programmed activities; who never felt the breath of the child psychologist on their neck. Because the early life of these children is a series of discoveries above and beyond play and work, their image of the world is compounded of their own observations and experiments. With Nature as their inexhaustible toy shop, children in primitive societies get far better ideas of how to have a good time than the most zealous kindergarten teacher. "The children in Uganda are very well behaved," noted an explorer; "I was struck by the way in which they amused themselves. Instead of making senseless mud pies, they made miniature villages which were almost exact copies of the dwellings around them. They would be thus employed for hours together, day after day, and would persevere until their models were completed."[20] There is no mention of an ensuing holocaust.

Among the Kpelle (the Liberian tribe whose communal spirit has been mentioned earlier) children are at liberty to do as they please. They are not at a loss of what to do; they build houses which they furnish with household things made by themselves; they dance, sing, and stage drum festivals; they cast magic spells and make offerings; they pretend getting engaged and married. Their power of observation is uncanny, their devotion to what they are doing total. Transition from play to work is imperceptible; when the children accompany their elders to the fields, they are given a small hoe and sickle to play with. It was Plato who counseled parents to "give real miniature tools to those three-year-olds." At all times, unspoiled children were builders at heart. In *The Clouds*, Stepsiades boasts to Socrates about his son Pheidippides, "when he was but a little child

306. Opposite page: Ethiopian children, innocent of mechanical toys and mechanized entertainment, their imagination unimpaired by pedagogics, amuse themselves with building models of huts and corrals.

he made a thousand pretty things, clay houses, castles, ships. . . ."

Not America but Africa emerges as the children's paradise on earth. "There is nothing more extraordinary," wrote a naturalist, Nathan Miller, on his visit to a Congo tribe, "than to see a young Muholoholo not only imitate all that his elders make—nets, mats, swords, paddles, spades—but to bring to them notable improvements."[21] In that sunny state of mind induced by unsolicited play, the child constructs new forms for baskets, evolves bizarre musical instruments, and invents traps for catching wild animals. In Africa and Oceania, wherever the native life style is still intact, children make their own tools. On Borneo and the Gilbert Islands they build their own houses and canoes. "The natural imitative skill of the primitive child knows no bounds," says Miller.[22] And yet, childishness remains unimpaired. Ethiopian children playfully put up small-scale huts and corrals, and stock them with snails, shells, and pebbles, that are their cattle, goats, and sheep. (By the same vivid stretch of imagination Japanese adults create *bonseki*, miniature landscapes, from stone and sand on lacquer trays, in which pebbles stand for sailing boats and flying geese. Unlike the children's ephemera, bonseki are treasured as family heirlooms.)

The children of the Warega, a Congolese tribe, do not only build entire model villages but act out an accelerated version of a full day's community life. Preparations begin on the eve. In the morning the children proceed to a river, catch fish, and roast them on fires in front of their own little houses. This done, they go about their work much like their parents. At a certain moment, one of them calls out, "It is night." Whereupon they retire and pretend to sleep. After a while another one imitates the crowing of a cock. They all now awake and recommence their work.[23] It is a far cry from our tutored tots, playing with blocks.

We can only guess at the thoughts the precocious young model builders might harbor should they happen to come across our peculiar nursery toys. A confrontation of some Warega children with, say, a missionary's block-addicted scion might tell us more about our educational follies than shelves of books. Would the little savages take to the slick cubes? If so, would their interest last? By the same token, would the nursery graduate be bright enough to join the impromptu play of the African children?

Culture is communicable intelligence, and play is at the very roots of culture. In a frigid environment, intelligence, the propelling power of creative activity, often defies efforts at its cultivation; temperature and temperament are not just etymologically linked. Nevertheless, even a cold climate need not deter children from playing outdoors.

"The building of snowhouses" wrote an optimistic Vilhjálmur Stefánsson, "could be taught by correspondence to boys in any place on earth where the winters are cold enough."[24] As an arctic explorer and builder of hundreds of snowhouses, he knew what he was talking about. "Building with snow blocks," he explained, "is far simpler than building with masonry." Snowhouses are indeed unique in that they require one building material only—which, like the biblical manna, falls from the sky—and a single working tool, a knife. A fastidious builder may also want a piece of string for tracing a perfect outline—"somewhat," Stefánsson suggested, "as a schoolboy may use two pencils and a string to make a circle."[25]

307. *Snow house at Takamachi, Niigata Prefecture, Japan.*

The procedure is simplicity itself. The snow is cut into slabs, 4 inches thick and 15 to 20 inches wide, and set up in an ascending spiral to form a domed roof. The chinks between them are filled with soft snow which hardens in ten minutes. Three practiced pairs of hands, Stefánsson maintained, can build a snowhouse 9 feet in diameter and 6 feet high in three-quarters of an hour. He certainly makes it sound like child's play. Besides, as a seasonal occupation it is infinitely more civilized than, say, snowmobile racing.

308. *At Sibiu, the former Hermannstadt, capital of Transylvania, children are sketching the ancient New Church's turreted tower. Such exercises are part of European school training that helps to make them aware of their country's architectural heritage.*

For some time these roots of culture have been endangered by a parasitic growth of toys—mostly objects without the redeeming grace of the homely and poetic playthings of former generations. Not that the traditional kind was always a challenge to a child's imagination; today's stultifying toys have their antecedents. The singular burst of building activity in the second half of the nineteenth century which metamorphosed Europe's towns in the days of Haussmann and during the Wilhelmian era also left its mark on the nursery. To keep the fires of block lust burning, children were given a *Steinbaukasten*—Germans have been all along the world's leading toy manufacturers—in English rendered, not very felicitously, as a "box of children's bricks." Its innovation consisted of new, ever more dulling elements: prefabricated architectural components such as columns, steeples, and arches of a strong Romanesque flavor. Whichever way one combined them, the result was a sort of embryonic Wartburg or Prince

Albert Memorial. The pseudo-Wagnerian and authentic Victorian tastes of the adults thus percolated to their children, introducing them painlessly to conspicuous inanity.

Until the Second World War, the box of bricks was the classical Christmas present. Richter and Co., the firm that dominated the juvenile building industry, even published a series of model designs that served as guidelines for the young master builders. Great must have been the chagrin of the parent whose distraught offspring failed to get the educational message and instead indulged in his private fantasies by perceiving the blocks as trees, animals, or people.

A far more deleterious plaything, the doll's house, offshoot of bourgeois opulence, had appeared on the domestic scene still earlier. Significantly, the place of this *jouet de luxe* was not in the nursery but in the parlor where it was exhibited for the admiration of guests

309. *German doll's house, dated 1631.*

rather than for the delectation of children. Too precious to be touched by clumsy little paws, its function was reduced to a showpiece. "After the sixteenth century," wrote a connoisseur of children's toys, "dolls houses were the common property of all European countries which could be reckoned as civilized."[26]

Our great-grandmothers imbibed their housewifely virtues and some of the tenets of the social code, together with deep draughts of class consciousness, while playing with dolls whose dishabillés included tight-laced stays, frilly drawers, and boots, or while being absorbed in the adoration of their miniature mansions. Playing with dolls Rilke thought to be the "initiation into the rigid passivity and emptiness of life."

In European countries, the unpretentious toymakers of old had been joined by craftsmen capable of reproducing the trappings of a well-run household on a minute scale—parquets and stuccoed ceilings, wallpaper and embroidered rugs, veneered tables and chests of drawers, chairs upholstered in silks and brocades, quilted bedspreads, glass chandeliers, marble fireplaces, and gilt-framed mirrors of Lilliputian proportions.

The insinuating monstrosities are far from extinct. Some of them found their way into museums; others, updated, are still current architectural/interior-decoratorish icons. "These houses," wrote a knowledgeable Englishman, "are microcosmic indications of the trends in domestic architecture for they have moved from the Edwardian red-brick dwelling house to the country cottage with roses around the door; from the suburban villa with side garage and red-tiled roof, to the detached spacious home in its own grounds; from the functional apartment block, with its streamlined and tubular furniture, to the maisonette."[27] One French toy, a villa of 1905 vintage, boasts of an elevator to the upper floor. Another, more sophisticated plaything (made for the English market) represents a *vespasienne*, the typical Parisian street urinal, with a man inside and a lady passing by.[28]

Lately children's bricks have surfaced in plastic and metal versions, conceived on the lines of contemporary assembly methods—slabs instead of stones, provided with slots, mortises, tenons, and of course with a generous amount of transparent components. All the affluent tot needs for his ultimate happiness are the modern equivalents of the Victorian dollhouse knick-knacks—washing machine and hibachi, electric typewriter and intercom, and in every nook a miniature color television set in good working order. Japanese toy manufacturers might oblige.

With these ready-made toys, the child picks up ready-made notions and acquires mental deformities that later put him in good standing

310. Detail, Victorian doll's house. (Courtesy Victoria and Albert Museum)

with society. The infantile paralysis of his creative faculties is of little moment. Trapped by television or intoxicated by machined music, with his young body and mind progressively immobilized, play becomes obsolete. If the parent could bring himself to desist from ingraining in the child's mind commercially conditioned precepts and prejudices, he might have to face fewer educational problems later on. If, as Maria Lluisa Borràs suggests, "he could manage to dispense with a whole international industry, then the child might be able to create his own toys, toys which would not aspire to be educational today, which might not be poetic tomorrow, but which the child would consider very much his own, something with the pulsing life of all that is self-created, imperfect and sufficient."[29] Alas, they don't make children that way anymore.

Note on "Architecture Without Architects"

The exhibition "Architecture Without Architects" which opened in 1964 at New York's Museum of Modern Art over the protests of the architectural fraternity had come about more or less inadvertently. In 1941, returning to New York after an absence of five years upon the museum's invitation, I was asked by Philip L. Goodwin, director of the Department of Architecture, to suggest some unhackneyed subjects for exhibitions. Vernacular architecture, I thought, would be a welcome departure from routine. Besides, ten years earlier, I had successfully tested its appeal by exhibiting my photo collection of anonymous architecture at the Bauausstellung in Berlin. In New York, however, the subject was considered unsuitable for a museum dedicated to modern art. If anything, it was thought to be antimodern. Nevertheless, a portfolio of my photographs was graciously accepted and buried in the museum files where it remained undisturbed for the next twenty-three years.

In 1960, the museum commissioned me to organize a number of didactic exhibitions, their subjects left to my discretion. This time vernacular architecture was found acceptable as a theme, provided the exhibition was going to be shown in out-of-town institutions only.

Vernacular architecture simply was not respectable. To make it so, it was desirable to win the token support of select colleagues, although their interest in tree houses and troglodytic dwellings could be expected to be no more than tepid. José Louis Sert, Gio Ponti, Kenzo Tange, and Richard Neutra responded amiably, while Walter

311. "Architecture Without Architects" at New York's Museum of Modern Art.

Gropius had to be gently persuaded to second a study project that was alien to his mind. The tide turned after Pietro Belluschi, dean of architecture at the Massachusetts Institute of Technology, saw the picture material. In a letter to the president of the John Simon Guggenheim Foundation, Gordon N. Ray, he wrote: "Somehow for the first time in my long career as an architect I had an exhilarating glimpse of architecture as a manifestation of the human spirit beyond style and fashion and, more importantly, beyond the narrows of our Roman-Greek tradition." A grant from the foundation made possible a field trip through eight countries—the twenty-fourth since my student days—which netted additional information and photographs.

At last, a showing of the exhibition on the Museum of Modern Art grounds was agreed upon, and a slender book, *Architecture Without Architects*, was hastily prepared in lieu of a catalog. Considered unsalable, it was intended as a giveaway to preferred museum members. Upon its publication, however, it was made generally available. Requests from museums at home and abroad resulted in circulating two identical travel editions. At the Louvre the title was bowdlerized to *Architectures méconnues, architects inconnus*—"Unrecognized architecture, unknown architects"—and on the whole the exhibition lived up to its reputation of being "truly subversive." In eleven years of travel the two editions were presented in eighty-four major museums and galleries from Australia all the way to behind the Iron Curtain. The museum's book, after running out of print one time too many, was entrusted to more discerning publishers.

Text references

Introduction with asides

1. Lucius Annaeus Seneca *Ad Lucilium epistulae morales* ep. 90.
2. *Encyclopaedia Britannica* (1947), 2:274.
3. Pliny the Elder *Natural History* (Cambridge, Mass., 1938–1962), 7.1.5–8.
4. Henri Bergson, *Creative Evolution* (New York, 1944), p. 167.

In praise of caves

1. Henri Breuil, *Four Hundred Centuries of Cave Art* (Montignac, 1952), p. 22.
2. Ibid., p. 23.
3. René Dubos, "Man Adapting: His Limitations and Potentialities," in *Environment for Man: The Next Fifty Years* (Bloomington, Ind., 1967), p. 13.
4. Porphyry, *On the Cave of the Nymphs in the Thirteenth Book of the Odyssey* (London, 1917), p. 11.
5. Bernard Rudofsky, *Behind the Picture Window* (New York, 1955), p. 22.
6. Paul Lucas, *Voyage du Sieur Paul Lucas, fait par ordre du Roy . . .* (Paris, 1712), 1:160.
7. Ibid., p. 161.
8. Ferdinand von Richthofen, *Tagebücher aus China* (Berlin, 1907), p. 535.
9. George B. Cressey, *Asia's Lands and People* (New York, 1951), p. 109.
10. Richthofen, *Tagebücher aus China*, p. 468.
11. George B. Cressey, *Land of the Five Hundred Million: A Geography of China* (New York, 1955), p. 263.

Brute architecture

1. *Michelin Paris* (1956), p. 177.
2. Tacitus *Germania* 46.
3. Sir Walter Raleigh, *The Discoverie of the Large, Rich and Beuutiful Empire of Guiana . . .* (London, 1569), p. 42.
4. A. H. Schultz, *The Life of Primates* (New York, 1969), p. 214.
5. Irven DeVore, *Primate Behavior* (Toronto, 1965), p. 449.
6. Ibid., p. 448.
7. Wolfgang Koehler, *The Mentality of Apes* (New York, 1927), p. 91.
8. Lewis H. Morgan, *The American Beaver* (Philadelphia, 1868), p. 83.

9. Charles Darwin, *The Origin of Species* (New York, Modern Library edition, 1948), p. 196.
10. Quoted in Remy Chauvin, *Animal Societies* (New York, 1968), p. 133.
11. J. P. Watson, "A Termite Mound in an Iron Age Burial Ground in Rhodesia," *Journal of Ecology* 55 (1967): 669.
12. Henry Smeathman, "Some Account of the Termites . . ." *Proceedings of the Royal Society of London* 71 (1781): 139.
13. P. E. Howse, *Termites: A Study in Social Behavior* (London, 1970), p. 81.
14. O. W. Richards, *The Social Insects* (London, 1970), p. 16.
15. Bergson, *Creative Evolution*, p. 173.
16. Maurice Maeterlinck, *Life of the White Ant* (London, 1927), p. 18.
17. Ibid., p. 8.
18. Jules Michelet, *L'Oiseau* (Paris, 1858), p. 208.
19. Gaston Bachelard, *La Poétique de l'espace* (Paris, 1957), p. 102.
20. René Dubos, *So Human an Animal* (New York, 1968), p. 16.
21. Gaston Bachelard, *The Poetics of Space* (New York, 1958), p. 93.
22. John Hurrell Crook, "A Comparative Analysis of Nest Structure in the Weaver Birds," *IBIS* 105:238.
23. John Hurrell Crook, "Nest Form and Construction in Certain West African Weaver Birds," *IBIS* 102: 11.
24. Ibid., p. 15.
25. J. Gould, *Handbook for the Birds of Australia* (London, 1855), 1:444.
26. E. S. Dixon, *The Dovecote and the Aviary* (London, 1851), p. 1.
27. Ibid., p. 2.
28. Ibid., p. 62.
29. Dorothy Calles, *Christian Symbols* (1971), p. 112.
30. Henry Maundrell, *A Journey from Aleppo to Jerusalem* (Oxford, 1732), p. 37.
31. Edmund Wilford, "Those Great Coops," *Architectural Review*, December 1962, p. 443.
32. Pliny the Elder *Natural History* 10.52.
33. James Hornell, "Egyptian and Medieval Pigeon-Houses," *Antiquity* (1947): 182.
34. Sir James Richards, "Bridegroom's Dowry," *Architectural Review*, February 1946, p. 60.
35. William J. Hamilton, *Researches in Asia Minor* . . . (London, 1842), 2:252.
36. Ibid., p. 255.

When architecture was all play and no work

1. Quoted in Glyn E. Daniel, *A Hundred Years of Archaeology* (London, 1950), p. 312.
2. Grahame Clark and Stuart Piggot, *Prehistoric Societies* (London, 1965), p. 290.
3. Inigo Jones, *The Most Notable Antiquity of Great Britain . . .* (London, 1725), p. 8.
4. A. P. Trotter, "Stonehenge as an Astronomical Instrument," *Antiquity* (1927): 43.
5. Quoted in Edgar Barklay, *Stonehenge and Its Earth-Works* (London, 1895), p. 143.
6. W. S. Blacket, *Researches Into the Lost Histories of America* (London, 1883), p. 193.
7. Ibid., p. 247.
8. Ibid.
9. Ibid., p. 193.
10. *Encyclopaedia Britannica* (1947), 2:91.
11. Hugh Henry Brackenridge, *Views Of Louisiana* (Baltimore, 1814), p. 187.
12. Ibid., p. 188.
13. Ibid., p. 182.
14. Quoted in Stephen D. Peet, *Prehistoric America* (Chicago, 1892), 1:162.
15. William Pidgeon, *Traditions of the De-coo-Dah* (New York, 1853), p. 19.
16. Ibid., p. 16.
17. Thomas Morton, *New English Canaan* (1637), p. 22.
18. Gerald S. Hawkins, *Stonehenge Decoded* (New York, 1965), book jacket.
19. Quoted in George E. Squier, *Peru Illustrated* (New York, 1877), pp. 468, 469.
20. William Atkinson, *Views of Picturesque Cottages . . .* (London, 1805), p. 13.
21. G. R. Kaye, *A Guide to the Old Observatories at Delhi* (Calcutta, 1920), p. 1.
22. Werner Buttler and Waldemar Haberey, *Die bandkeramische Ansiedlung bei Köln-Lindenthal* (Berlin, 1936), p. 36.
23. Gaston Bachelard, *The Poetics of Space*, p. xxxiv.
24. Buttler and Haberey, *Die bandkeramische Ansiedlung*, p. 34.
25. *Bretagne* (Paris, 1973), p. 574.
26. *Encyclopaedia Britannica* (1947), 4:890.

27. *Brittany* (Paris, 1971), p. 26.

28. *Bede's Ecclesiastical History of the English People* (London, 1969), p. 107.

29. N. K. Sandars, *Prehistoric Art in Europe* (Harmondsworth, England, 1968), p. 218.

30. Julio Martinez Santa-Olalla, "The Cyclopean Walls of Tarragona," *Antiquity* (1936): 75.

Mobile architecture

1. Herodotus *Historiae* 4.46.

2. Hippocrates *Airs, Waters, and Places* (London, 1881) 6, p. 93.

3. Pierre de Bergeron, *Relation des Voyages en Tartarie* (Paris, 1634), p. 12.

4. Ibid., p. 14.

5. Pietro della Valle, *Viaggi di Pietro della Valle* (Rome, 1658), 2:487.

6. David Livingstone, *Missionary travels* (London, 1857), p. 286.

7. Quoted in ibid., p. 391.

8. Herodotus *Historiae* 4.23.

9. Arthur Upham Pope, *Survey of Persian Art* (London, 1939), 2:1419.

10. Ibid., p. 1411.

11. *The Travels of Marco Polo* (Rochester, N.Y., 1933), p. 179.

12. Ibid.

13. Edgar Blochet, *Musulman Painting* (London, 1929), plate lxxxvi.

14. *Amerikanisches Magazin* 1 (Hamburg, 1929): 137.

15. Johan Nieuhof, *An Embassy from the East-India Company* (London, 1669), p. 95.

16. Joseph Needham, *Science and Civilization in China* (Cambridge, England, 1956–1971), vol. 4, part 3, p. 394.

17. Nieuhof, *East-India Company*, p. 95.

18. Ibid.

19. Needham, *Science and Civilization in China*, vol. 4, part 3, p. 694.

20. Erasmus Francisci, *Ost- und West-Indischer Lustgarten* (Nürnberg, 1668), p. 129b.

21. *Ante-Nicene Fathers: Origen against Celsus* (New York, 1917–1925), book 4, p. 516.

22. Don Cameron Allen, "The Legend of Noah," *Studies in Language and Literature* (Champaign/Urbana, Ill., 1949), p. 81.

Storehouses, sepulchral and cereal

1. Sir Thomas Browne, *Urne Buriall* (Cambridge, England, 1958), p. 23.

2. Pedro Cieza de Léon, *The Incas* (Norman, Okla., 1959), p. 274.

3. W. H. Dali, "Notes on Pre-Historic Remains . . . ," *Proceedings of the California Academy of Sciences,* November 4, 1872, p. 2.

4. John Lubbock, *Pre-Historic Times* (New York, 1913), p. 508.

5. Sir William M. Flinders Petrie, *Gizeh and Rifeh* (London, 1907), p. 14.

6. N. de Garis Davies, "The Town House in Ancient Egypt," *New York Metropolitan Museum Studies* 1 (1928–29): 250.

7. Marcus Vitruvius Pollio, *De Architectura* (London, 1874) 2.8.20.

8. Pliny the Elder *Natural History* 10.73.

9. Machteld J. Mellink, "The Early Bronze Age in Southwest Anatolia," *Archaeology* 22 (October 1969): 291.

10. Ibid.

Strongholds

1. Marcel Griaule and Germaine Dieterlen, "The Dogon," in *African Worlds* (London: International African Institute, 1954), p. 97.

2. Sigmund Freud, *Civilization and Its Discontent* (New York, 1958), p. 38.

3. *Memoirs of the Berenice Pauahi Bishop Museum* (Honolulu, 1906–1909), 2:243.

4. Edward Lear, *Journals of a Landscape Painter in Southern Calabria . . .* (London, 1852), p. 35.

5. Ludwig Ross, *Reisen auf den griechischen Inseln . . .* (Halle a. S., 1845), Lieferung 20–23, p. 53.

6. Bernhard Rudofsky, *Eine primitive Betonbauweise auf den südlichen Kykladen . . .* (doctoral dissertation, Vienna, 1931), p. 21.

7. Ernle D. S. Bradford, *The Companion Guide to the Greek Islands* (New York, 1963), p. 145.

8. Michel Santiago, "Functional Tradition," *Architectural Review,* March 1960, p. 208.

The vanishing vernacular

1. R. Curzon (Zouche), *Visits to the Monasteries in the Levant* (London, 1865), p. 372.

2. Quoted in Esther Singleton, *Romantic Castles and Palaces* (New York, 1901), p. 347.

3. Daniel Speckle, *Architectura von Vestungen* (Strasbourg, 1589).

4. Bernard Berenson, *Aesthetics and History* (New York, 1948), p. 93.

5. Johann Gustav Gottlieb Büsching, *Ritterzeit und Ritterwesen* (Leipzig, 1823), 1:135.

6. Ibid., p. 138.

7. Ulrich von Lichtenstein, *Der Frauendienst* (Stuttgart, 1924), p. 247.

8. Michio Taketama, "Rebirth of the Castles," *This Is Japan* (Tokyo, 1961), p. 107.

9. Marina and Robert Adams, "Santorini Reborn," *Architectural Review*, May 1961, p. 372.

10. Moshe Safdie, *Beyond Habitat* (Cambridge, Mass., 1970), p. 117.

11. Lancelot G. Bark, "Beehive Dwellings of Apulia," *Antiquity* (1932): 407.

12. Quoted in Alice Brayton, G. *Berkeley in Apulia* (Boston, 1946), p. 51.

The vernacular obliquely appraised

1. Bill Kovach, "Struggle for Identity," *New York Times*, November 27, 1970.

2. Stewart L. Udall, *The Quiet Crisis* (New York, 1963), p. 55.

3. S. F. Markham, *Climate and the Energy of Nations* (New York, 1944), p. 160.

4. Quoted in Oscar Newman, *CIAM '59 in Otterlo* (1961), p. 213.

5. Diedrich Westermann, *Die Kpelle: Ein Negerstamm in Liberia* (Göttingen, 1921), p. 73.

6. Charles Wilkes, "Narrative of the U.S. Exploring Expedition During the years 1838–42," *Memoirs of the Berenice Pauahi Museum* (Honolulu, 1906–1909), 3:210.

7. John Murray, *A Handbook for Travellers in Spain and Readers at Home* (London, 1845), p. 556.

8. D. H. Lawrence, *Twilight in Italy* (London, 1916), p. 280.

9. Ibid., p. 281.

10. Ibid., p. 282.

11. Arthur Mitchell, *The Past in the Present* (Edinburgh, 1880), p. 62.

12. Ibid., p. 64.

13. Ibid., p. 51.

14. Ibid., p. 52.

15. Ibid., p. 55.

16. John Ruskin, *The Poetry of Architecture* (Sunnyside, Orpington, England, 1893), p. 10.

17. Ibid., p. 18.

18. Ibid., p. 22.

19. Ibid., p. 12.

20. André Gide, *Travels in the Congo* (New York, 1930), p. 217.

21. Ibid., p. 218.

22. Vitruvius *De Architectura* 3.1.4.

23. Ibid., 6.6.1.

24. Atkinson, *Views of Picturesque Cottages . . .* , chap. 5.

25. Ibid., chap. 6.

26. Henry David Thoreau, *Walden* (New York, 1938), p. 37.
27. Ibid., p. 36.
28. Ibid., p. 206.

The importance of trivia

1. Quoted in Wilkes, "Narrative of the U.S. Expedition," p. 192.
2. Victor W. von Hagen, *The Desert Kingdom of Peru* (New York, 1964), p. 57.
3. A. L. Sadler, *A Short History of Japanese Architecture* (Sidney, 1941), p. 98.
4. Father Gramont, "An Account of the K'ang, or Chinese Stoves," *Philosophical Transactions Abridged* 61 (London, 1809): 95.
5. *Encyclopaedia Britannica* (1947), 3:294c.
6. Ibid., 20:726d.
7. Joseph Pitton de Tournefort, *A Voyage Into the Levant* (London, 1741), 2:332.
8. James Mellaart, *Çatal Hüyük* (New York, 1967), p. 60.
9. Quoted in Peter Collins, *Changing Ideals in Modern Architecture* (London, 1965), p. 24.

The call of the Minotaur

1. Sigfried Giedion, *The Eternal Present* (New York, 1964), p. 495.
2. Tournefort, *A Voyage Into the Levant,* 1:65.
3. Ibid., p. 67.
4. F. W. Sieber, *Reise nach der Insel Kreta* (Leipzig, 1823), 1:515.
5. Georg Wilhelm Friedrich Hegel, *Vorlesungen über die Aesthetik* (Berlin, 1928), 2:287.
6. Stephen Jones, *History of Poland* (London, 1795), p. 29.
7. Ibid., p. 30.
8. Ibid., p. 32.
9. Ibid., p. 31.

Block lust

1. J. C. Friedrich von Schiller, *On the Aesthetic Education of Man* (London, 1964), fourteenth letter, p. 79.
2. T. S. Eliot, "Notes Towards a Definition of Culture," *Partisan Review,* (1944): 148.
3. *Encyclopaedia Britannica* (1947), 18:72b.
4. Friedrich Wilhelm Fröbel, *The Education of Man* (New York, 1885), p. 31.
5. Harriet Merril Johnson, *The Art of Block Building* (New York, 1933), p. 24.

6. Ruth E. Hartley et al., *Understanding Children's Play* (New York, 1952), p. 99.

7. Arnold Gesell et al., *The First Five Years of Life* (New York: Harper & Row; London: Methuen, 1940), p. 109.

8. Hartley et al., *Understanding Children's Play*, p. 99.

9. Johnson, *Art of Block Building*, p. 6.

10. Emil Menzel, "Patterns of Responsiveness in Chimpanzees," *Psychologische Forschungen* 27 (Vienna, 1964): 337.

11. Gesell et al., *First Five Years*, p. 110.

12. Ibid., p. 114.

13. Ibid., p. 109.

14. Frances Guanella, "Block Building Activities of Young Children," *Archives of Psychology*, no. 174 (1934): 8.

15. Hartley et al., *Understanding Children's Play*, p. 111.

16. Ibid., p. 115.

17. Gesell et al., *First Five Years*, p. 113.

18. Guanella, "Block Building Activities," p. 22.

19. Desmond Morris, *The Biology of Art* (London, 1962), p. 13.

20. Robert W. Felkin, "Notes on the Waganda Tribe of Central Africa," *Proceedings of the Royal Society of Edinburgh* 13 (1884–1886): 745.

21. Nathan Miller, *The Child in Primitive Society* (New York, 1928), p. 142.

22. Ibid., p. 141.

23. Charles Delhaise, *Les Warega* (Brussels, 1909), p. 67.

24. Vilhjálmur Stefánsson, *The Friendly Arctic* (New York, 1943), p. 175.

25. Ibid., p. 174.

26. Karl Gröber, *Children's Toys of Bygone Days* (London, 1932), p. 22.

27. Leslie Daiken, *Children's Toys* (New York, 1953), p. 126.

28. Jac Remise and Jean Fondin, *The Golden Age of Toys* (Chicago, 1967), p. 39.

29. Maria Lluisa Borràs, *Mundo de los Juguetes* (1969), p. 34.

Index

Acknowledgments

Three grants from the John Simon Guggenheim Memorial Foundation made possible studies on four continents in preparation for this and a preceding book on the same subject. The archaeologists H. Džambov, Waldemar Haberey, Machteld J. Mellink, Pierre Demargne, and Jan Zahle obliged me with valuable information, as did Don Tomás Zueco, Mayor of Tarazona. Sincere thanks go to Ruth M. Anderson, the Hispanic Society of America; D. E. Dean, British Architectural Library; Jack Jacoby and Philip Yampolsky, Columbia East Asian Library; Kazimierz Salewicz, National Museums, Copenhagen; above all, to Elizabeth Roth, Prints Division, New York Public Library. Among professional and nonprofessional photographers who generously made available their work are to be named Roberto Brambilla, Wulf Dieter Graf zu Castell-Rüdenhausen, William R. Current, William O. Field, Yukio Futagawa, Hiroshi Hamaya, Evelyn Hofer, Gunda Holzmeister, Saara Hopea-Untracht, Mian Abdul Majid, Raphaël G. Mischkind, José Moreno Naranjo, Isamu Noguchi, José Ortiz Echagüe, Torao Saito, Sigrid Spaeth, and Vilko Zuber. The various archives, libraries, museums, galleries, government agencies, and individuals who furnished pictorial material are listed either in the picture captions or in the Sources of Illustrations.

I also wish to thank the staff of the publishers, particularly Daniel Okrent, editor-in-chief, Kay Lee, untiring counselor and cheerful helper, and Bill Green, manuscript and production editor. As always, my greatest indebtedness is to my wife, assistant and critic.

Sources of illustrations

Accademia di Belle Arti, Venice/Rossi 246. Alinari 80, 139. American Museum of Natural History 61. Roberto Brambilla 206, 210. Museo Canario, Las Palmas 9, 132. Wulf Dieter Graf zu Castell-Rüdenhausen 25. Columbia East Asian Library 2, 27, 35, 102, 107. William R. Current 6. Danish National Museums 137, 164. Pierre Demargne 159. M. G. J. Duchemin 269. Ente Nazionale Industrie Turistiche 84, 87, 217. Ente Provinciale per il Turismo, Treviso/Mazzotti 29. Hassan Fathy 243, 275. William O. Field 183. Fogg Art Museum 10. French National Tourist Office 75. Frobenius Institut, Frankfurt am Main 131, 151, 153, 200, 306. Yukio Futagawa 221, 230, 259. Gabinetto Fotografico Nazionale, Rome 223, 291. René Gardi 150. German Archaeological Institute, Athens 108. Greek National Tourist Office 196. A. Guillen 241. Hiroshi Hamaya 307. Hispanic Society of America, New York 149, 190. Evelyn Hofer 168. Gunda Holzmeister 16. Saara Hopca-Untracht 48. Martin Hürlimann 240. Léni Iselin 41. A. F. Kersting 95. Kjeld Kjeldsen 38, 152. Malta Government Tourist Office 91. Office Marocain du Tourism 244. MAS, Barcelona 74, 81, 224. Raphaël G. Mischkind 222. Isamu Noguchi 64. José Ortiz Echagüe 192. Pressehuset, Copenhagen/Betting 46. Bernard Rudofsky 1, 5, 11, 14, 15, 18, 19, 22, 23, 24, 26, 52, 54, 63 (p. 99), 65, 66, 72, 73, 77, 89, 90, 93, 98, 113, 129, 130, 134, 144, 145, 146, 147, 148, 154, 157, 161, 163, 166, 173, 174, 175, 177, 179, 181, 185, 187, 188, 193, 194, 195, 197, 198, 201, 202, 203, 204, 211, 215, 218, 220, 229, 231, 232, 233, 234, 237, 245, 248, 250, 251, 252, 253, 255, 256, 258, 264, 267, 268, 270, 271, 272, 273, 274, 276, 277, 278, 279, 280, 281, 282, 283, 284, 285, 286, 287, 290, 297, 299, 300, 302, 303, 308, 311. Torao Saito 191. Sindacato Nacional dos Arquitectos, Lisbon 112, 158, 227, 228. Sigrid Spaeth 121, 122. Spanish National Tourist Office 156. Joseph J. Stefanelli 51. Swiss National Tourist Office 155. Touring Club Italiano 212. Turkish Government Tourist Office 20, 21, 53, 55. United Nations/Rothstein 62. Museum für Urgeschichte, Niederösterreich 79. Juan Vidal Ventosa 47. Victoria and Albert Museum 162, 310. Vilko Zuber, Zagreb 301.